# Computer Mapping of Natural Resources and the Environment

## Including Applications of Satellite-Derived Data

## Volume Ten

Patricia A. Moore, Editor

VE RI TAS

HARVARD LIBRARY OF
COMPUTER GRAPHICS/1980
MAPPING COLLECTION
HARVARD UNIVERSITY

LABORATORY FOR COMPUTER
GRAPHICS AND SPATIAL ANALYSIS
520 GUND HALL, 48 QUINCY STREET
CAMBRIDGE, MASSACHUSETTS 02138

**ISBN 0-89866-019-X**

# Contents



*[Continued over page]*

# The Design and Applications of the LANDSAT-D Assessment System

*by P. A. Bracken, G. G. Knoble, C. B. Howell, and P. H. Van Wie*

## INTRODUCTION

The National Aeronautics and Space Administration initiated the LANDSAT-D research and development program in September 1977. The program includes launching an observatory in late 1981 which will incorporate a Thematic Mapper (TM) and a Multispectral Scanner (MSS). The spacecraft is designed to be compatible with the Space (Shuttle) Transportation System's satellite retrieval and replacement capabilities. The LANDSAT-D spacecraft and sensors will be operated through a Tracking and Data Relay Satellite (TDRS) System to produce an average of 50 TM and 200 MSS scenes per day over selected regions of the Earth.

A mini-computer-based LANDSAT-D Assessment System (LAS) is being developed at the Goddard Space Flight Center as part of the LANDSAT-D ground segment. The intent of the LAS is to assess the characteristics and utility of the Earth observations provided by the LANDSAT-D system, particularly the Thematic Mapper. The LAS will be used to process, display, analyze, and evaluate LANDSAT-D data in support of remote-sensing demonstrations for a variety of Earth resources applications.

## LANDSAT-D PROGRAM OBJECTIVES

The four major objectives of the LANDSAT-D program are:

- To assess the capability of the Thematic Mapper to provide improved information for Earth resources management.
- To provide for system-level feasibility demonstrations in concert with user agencies in order to define the need for and characteristics of an operational system.
- To encourage foreign participation in the program.
- To provide a transition for both domestic and foreign users from the Multispectral Scanner data to the higher resolution and data rate of the Thematic Mapper.

The purpose of the LAS and associated LANDSAT-D assessment activities is to demonstrate by analysis and test that the LANDSAT-D System meets the program objectives. To accomplish this, the LAS will be organized to support the assessment activities of investigators in at least five major disciplines: agriculture, hydrology, land use, forestry/rangeland, and geology. The total LAS effort will have the objective of providing insight and experience that will assist user agencies and all other parties interested in applying the LANDSAT-D data.

## LANDSAT-D SYSTEM OVERVIEW

The overall LANDSAT-D system is shown in Figure 1. The Operations Control Center (OCC), the Data Management System (DMS), the Domestic communications satellite (Domsat) Station and the Transportable Ground Station as shown in the Figure are all located at the Goddard Space Flight Center.

The LANDSAT-D data will be received at Goddard directly through the Transportable Ground Station or through a TDRS/Domsat link. The latter data path uses a TDRS ground station at White Sands, New Mexico, for relaying the data through Domsat to Goddard, and serves as the main data acquisition link for the LANDSAT-D ground data processing system.

The Department of the Interior's EROS Data Center will receive processed LANDSAT-D data in digital and film product form from Goddard and will maintain data archives. This agency will also assume responsibility for disseminating LANDSAT-D data products to the public. Other domestic and foreign users may enter into agreements with NASA to utilize their own ground stations to acquire the data directly from the LANDSAT-D satellite.

### LANDSAT-D Flight Segment

The LANDSAT-D flight segment includes an observatory which contains the Thematic Mapper and Multispectral Scanner instruments. The observatory will be launched into a medium altitude, sun-synchronous orbit in the third quarter of calendar year 1981. Table I describes the characteristics of the currently planned LANDSAT-D orbit.

Table I
Currently Planned LANDSAT-D Orbit

| | |
|---|---|
| Altitude | 705 Km |
| Inclination | 98.2 degrees |
| Sun Angle | 9:30 a.m. at descending node |
| Coverage Cycle | 16 days |

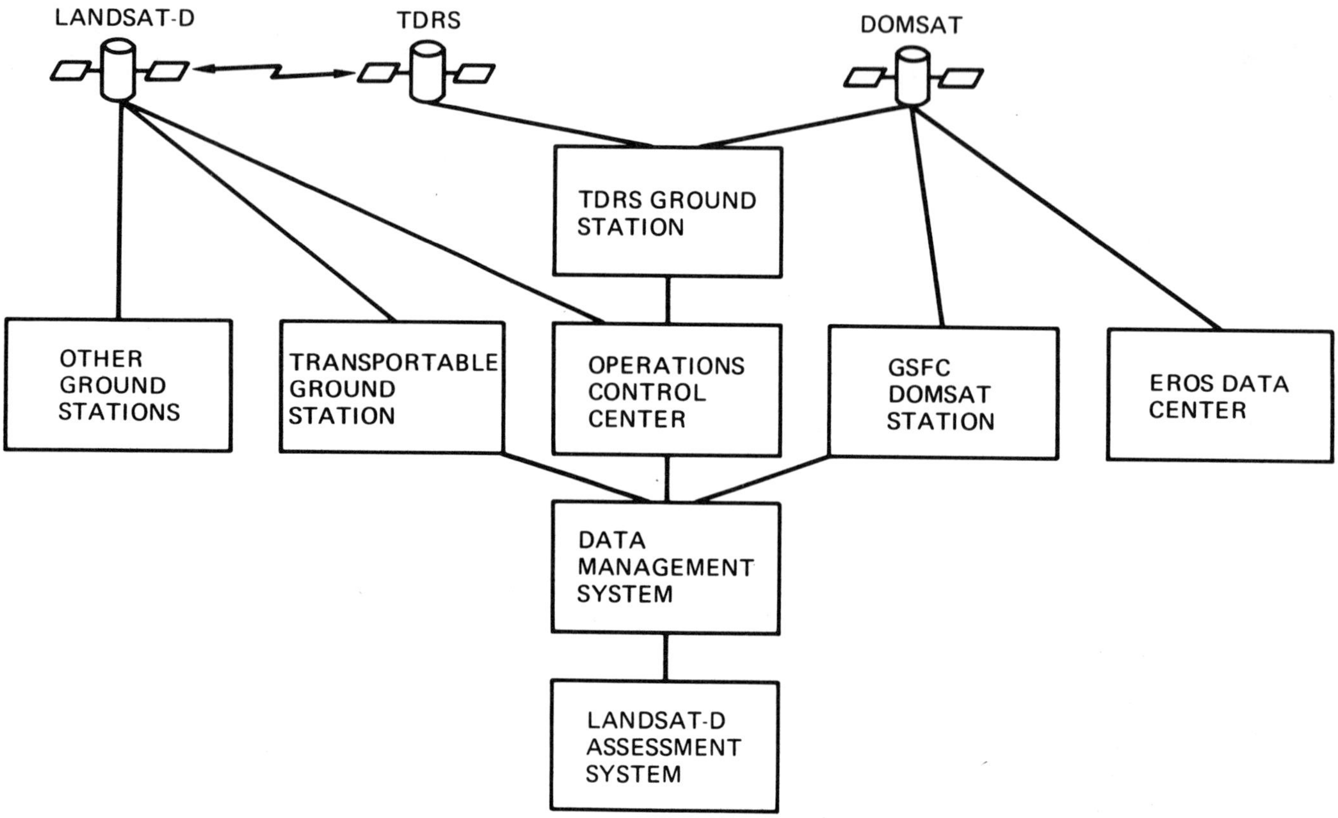


*Figure 1. The LANDSAT-D System*

Salomonson and Hall[5] describe several alternative orbits still under consideration. The final orbit selection will be made by late summer 1979.

The LANDSAT-D observatory is designed to incorporate the Multimission Modular Spacecraft which provides the modularized and standardized power, propulsion, attitude control, and communications and data handling subsystems needed for mission operations. The mission-unique subsystems designed into the spacecraft include the instruments, the solar array subsystem, a TDRS antenna, a wideband communications module, and a Global Positioning System[3] to support LANDSAT-D position determination experiments.

Table II lists the characteristics of the Thematic Mapper and Multispectral Scanner instruments.

The Thematic Mapper provides higher spatial and spectral resolution, additional spectral channels, higher signal-to-noise ratio, and greater radiometric and geometric accuracy than the LANDSAT Multispectral Scanner instruments. These improvements will provide greater accuracy in identifying, classifying, and measuring Earth observation parameters. The TM characteristics listed in Table II should improve Earth resources monitoring capabilities for a variety of disciplines, but were chosen to favor applications in agriculture, forestry, and rangeland. Blanchard and Weinstein[1] discuss the design and development of the Thematic Mapper instrument.

Table II
LANDSAT-D Sensor Characteristics

| | TM | MSS |
|---|---|---|
| Spectral Channels | 0.45-0.52 m | — |
| | 0.52-0.60 | 0.5-0.6 m |
| | 0.63-0.69 | 0.6-0.7 |
| | 0.76-0.90 | 0.7-0.8 |
| | 1.55-1.75 | 0.8-1.1 |
| | 2.08-2.35 | — |
| | 10.40-12.50 | — |
| Ground IFOV | 30m (Channels 1-6) | 80m |
| | 120m (Channel 7) | — |
| Ground Swath | 185KM | 185KM |
| Data Rate | 85 Megabits/sec | 15Mbs |
| Quantization Levels | 256 | 64 |

### LANDSAT-D Ground Segment

The LANDSAT-D ground segment is being assembled for NASA by the General Electric (GE) Company's Space Division. The major elements of the ground segment include the Transportable Ground Station, the Operations Control Center, the Data Management System, and the LANDSAT-D Assessment System.

The Transportable Ground Station will be used to acquire the LANDSAT-D data directly for engineering evaluation of the spacecraft and sensors and will serve as a backup data acquisition system in case of failure in the satellite data relay communications links.

The Operations Control Center will coordinate scheduling of the ground system resources and will provide the satellite command and control functions required to support the mission.

The Data Management System is the heart of the LANDSAT-D ground data processing system. As shown in Figure 2, the system will perform the digital image-processing functions required to produce corrected Thematic Mapper and Multispectral Scanner data products. The system is designed to process and correct up to 50 TM and 200 MSS scenes per day. A key performance requirement for the system is to produce final output products within 48 hours of data acquisition at Goddard. Smith and Webb[6] describe the Data Management System and its processing capabilities.

The functions and capabilities of the LANDSAT-D Assessment System are discussed in the next two sections.

## LANDSAT-D ASSESSMENT SYSTEM TASKS

The purpose of the LANDSAT-D Assessment System is to demonstrate by analysis and test that the LANDSAT-D System meets the objectives of the LANDSAT-D Program. The LAS will be used by investigator teams in cooperation with the project scientists to analyze and verify the LANDSAT-D output products and to develop improved means of processing the data for a variety of applications. The kinds of LAS tasks envisioned to accomplish the required support include:

Calibration and noise removal in TM data.
Position determination and image registration.
Optimizing data processing procedures and algorithms.
Assessing radiometric sensitivity, frequency of coverage, and data compression.
Assessing the benefits from improved TM resolution.
Support system level applications tests and demonstrations.

The above tasks will be performed to refine the data-processing and analysis procedures used with the TM data, to evaluate the incremental improvement offered by the Thematic Mapper over the Multispectral Scanner, and to improve the ability of users to apply the data in representative Earth observations applications.

## LAS - SYSTEM DESIGN

A block diagram of the LANDSAT-D Assessment System is shown in Figure 3. The system includes a Digital Equipment Corporation VAX-11/780 mini-computer with one megabyte of main memory, seven high-speed input/output ports, a low-speed input/output bus for connecting standard computer peripherals, a Floating Point System Ap-120B pipelined processor, and a variety of other peripheral devices discussed in the next section.

The inputs to the system are TM and MSS data from com-

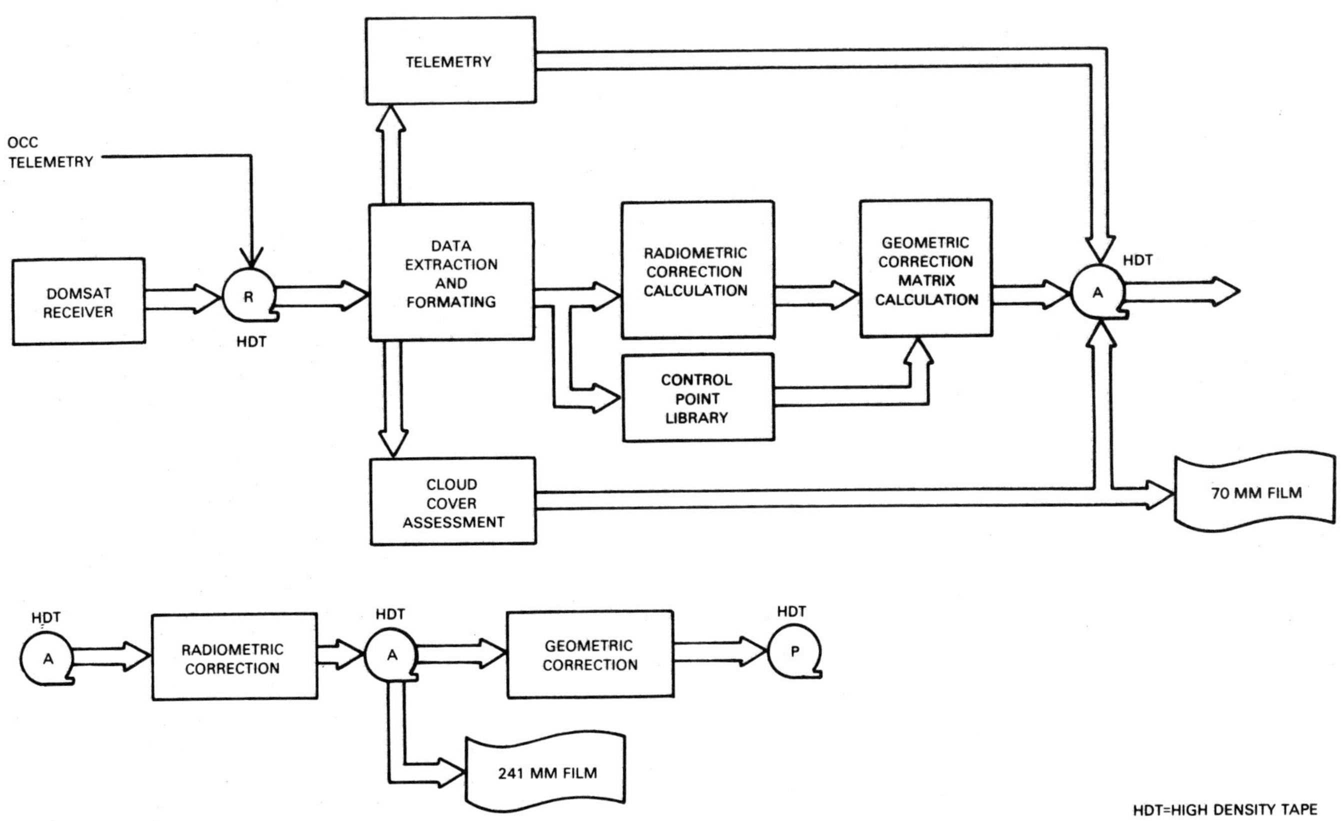


*Figure 2. LANDSAT-D Data Management-System Product Generation HDT - High density tape*

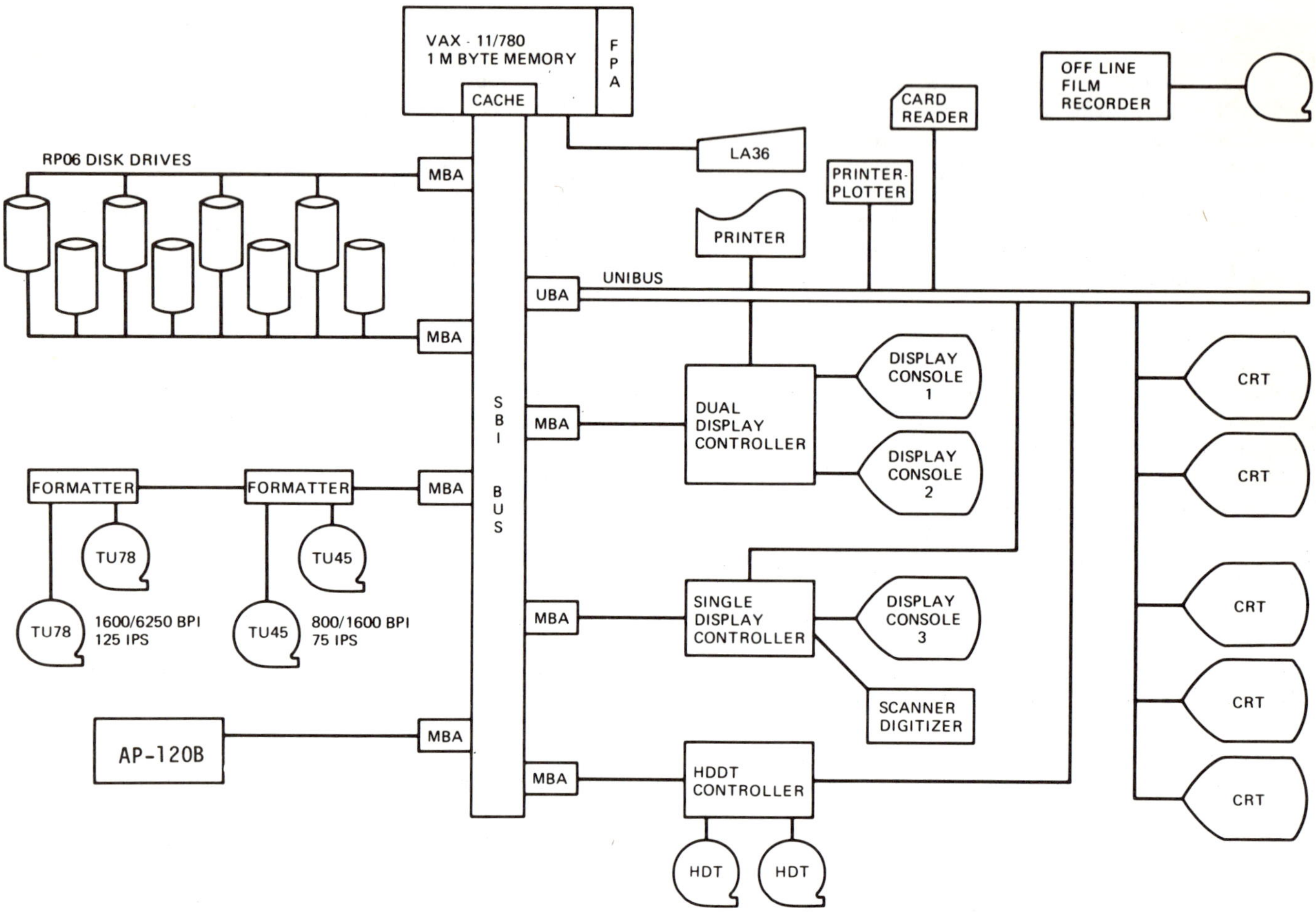


*Figure 3. LANDSAT-D Assessment System Configuration*

puter-compatible tapes and high-density tapes as well as ancillary data from tape, film, maps, charts, and punched cards. The primary interface between the Data Management System and the LANDSAT-D Assessment System will consist of 28 track high-density digital tapes containing radiometrically corrected data in band interleaved by line format, and radiometrically and geometrically corrected data in band sequential format.

The LAS has been sized to process image segments varying from 512 lines by 512 picture elements per line to full Thematic Mapper scenes of approximately 6500 lines by 6500 picture elements per line. Due to the volume of TM data, LAS capabilities for extensive processing of full TM scenes are limited. The expected daily input scene-processing workload for the LAS is listed in Table III.

Several additional assumptions were used in sizing the system:

The required image segments listed in Table III would be extracted from up to 12 (24 for the peak load) separate input scenes.

Thirty percent of the selected input data sets are required within 48 hours of data acquisition and the remaining data sets within two weeks of acquisition.

The typical LAS user session would involve extracting selected image segments, displaying the segments, and performing a variety of analysis operations such as multispectral classification and applications modeling.

Table III
LAS Daily Scene-Processing Load

| Image Segment Size | Number of TM Segments | Number of MSS Segments |
|---|---|---|
| 4000 lines x 4000 pixels | 1 | 1 |
| 2000 x 2000 | 4 | 4 |
| 1000 x 1000 | 3 | 3 |
| 512 x 512 | 4 | 4 |

TOTAL TYPICAL WORKLOAD — 350 megabytes/day
PEAK LOAD — 700 megabytes/day

Using these assumptions, typical LAS operations scenarios and processing timelines were developed, and a computer simulation was conducted by GE personnel to verify the throughput capabilities of the system configuration. The simulation verified that the LAS configuration has the capacity to meet the projected processing timelines required for ingesting, processing, displaying, and analyzing the LANDSAT-D data in a two-shift, 16 hour per day operation.

The hardware shown in Figure 3 and the required operating

system software are being implemented by GE and will be installed and operational in October 1980 — well before the launch of LANDSAT-D in late 1981. Required applications software is being developed by Goddard in parallel with GE's efforts and will be installed on the LAS starting in October 1980.

### LAS Hardware Characteristics

As shown in Figure 3, the LAS is configured around a VAX-11/780 mini-computer. The VAX computer includes a low-speed bus, the UNIBUS, and a high-speed mass bus, the Serial Backplane Interconnect (SBI) Bus. A Versatec printer/plotter, a 300 line per minute line printer, a card reader and five DEC model VT100 alphanumeric terminals are interfaced to the system through the UNIBUS.

The 1600/6250 bits per inch (bpi) magnetic tape drives and two 800/1600 bpi tape drives are connected to the mass bus and provide capabilities for reading and writing computer-compatible tapes. These units will serve for ingesting and creating data tapes and for creating tapes to be used as input to the off-line film recorder subsystem.

Two 28 track Martin Marietta high-density tape drives will provide the primary means for ingesting and archiving the LANDSAT-D data. The high-density tape drives record data on up to 24 data tracks at 20,000 bits per inch per data track. The remaining tracks are used for tape synchronization, a time code, an optional audio track and a spare track.

The high-density tape controller provides capabilities for searching tapes for a designated data set at 240 inches per second (ips) and for rewinding at 240 ips. These units will be operated at data transfers rates of 100 to 600 kilobytes per second on the LAS.

On the average, LAS investigators will require daily scene segments from up to 24 full TM and MSS scenes, possibly residing on four separate high-density tapes. Each high-density tape can contain up to 25 full TM or 50 full MSS scenes.

The high-density tape drive controller includes hardware to perform image cropping, subsampling, and/or averaging operations. Using these capabilities, overviews of a full scene and selected subscenes can be generated by the controller prior to transferring data to the VAX computer. This feature will greatly reduce the volume of data handled by the LAS during data ingest operations and will reduce system input/output time-lines from 17 to 5 hours per day to meet the expected daily input requirements listed in Table 3.

The LAS configuration includes eight RP06 disk drives — four of them dual-ported to allow multiple access paths and each providing 176 megabytes of storage. The four dual-ported disk drives provide the capacity to store up to two full TM and two full MSS scenes. These units will be used for bulk storage of full and partial scenes read from high-density tapes.

Of the remaining four disk drives, one will be used as the system disk and three will be dedicated to the three interactive image analysis consoles. Since the disk packs are removable on all eight drives, users will be assigned their own disks to store data. The use of removable disks also allows rapid mounting or dismounting of disks at the beginning and end of an analysis session.

The three LAS image display and analysis consoles will be developed by the Hazeltine Corporation and are similar to existing image analysis terminals already in use at the Goddard Space Flight Center.[2] The three consoles provide fully interactive image processing and color image display capabilities. Each terminal includes eight solid-state TV refresh memories with individual refresh channels containing 512 x 512 eight bit picture elements. The terminals will be interfaced to a VAX computer through a high-speed mass bus adapter, allowing data transfer of one 512 x 512 eight bit image in less than one second. Fast image manipulation is accomplished by the use of terminal hardware including a joystick driven cursor, image shifting and windowing circuitry, gray level look-up tables, and a switchable matrix memory configuration. System flexibility is maintained by controlling all hardware functions from the terminal driver software.

To supplement the processing capabilities of the VAX mini-computer, the LAS has been configured to include a Floating Point Systems AP-120B pipelined processor. This unit is capable of performing up to 12 million floating point operations per second using 38 bit floating point operands. The LAS AP-120B includes 256K words of data memory, 8K words of program memory, and 8K words of random access table memory. It is expected to speed up processing times by a factor of two to five over the VAX for geometric correction/image resampling, maximum likelihood classification, feature selection, mapping transformations, and other image-processing functions.

A block diagram of the off-line LAS film recorder subsystem is shown in Figure 4. This subsystem consists of a DEC PDP-11/34 mini-computer, an Optronics L5500 film recorder, a Dicomed D47 film recorder, an 800/1600 bpi tape drive, and a 1600/6250 bpi tape drive. The film recorders and tape drives will all be connected to the 11/34's UNIBUS, and be capable of simultaneous operation. The Optronics recorder will be used to produce high quality 241 mm square, full TM resolution, color separation negatives. The Dicomed will be used to produce film products of less than 4K x 4K resolution, such as TM subscenes and up to a full MSS scene, on a single color negative, or black and white or color polaroids. It is expected that four to six color products will be generated by the Optronics recorder and some combination of 10 to 20 products by the Dicomed per day.

To supplement the outputs of the image recorders, a Dunn Instruments 631 color camera system will be used to produce quick look, 8″ x 10″ color polaroids from the interactive displays. The camera will be connected to the video outputs of the image analysis terminals by a video switch to allow pictures to be taken from any of the 24 TV refresh memories.

### LAS Software

Since the intent of the LAS is to support teams of investigators in a variety of Earth resources applications, the LAS software is being designed to provide a user-oriented, interactive, and batch processing system structured for flexibility and rapid response to user requests. There are three main components of the LAS software: the DEC supplied VAX/VMS operating system, an LAS Executive, and the specific applications support software packages required by the investigator teams.

The VMS software is a multiple user, virtual memory operating system with automatic program paging features. VMS schedules and allocates system resources and includes a FORTRAN IV PLUS compiler, a MACRO assembler and ex-

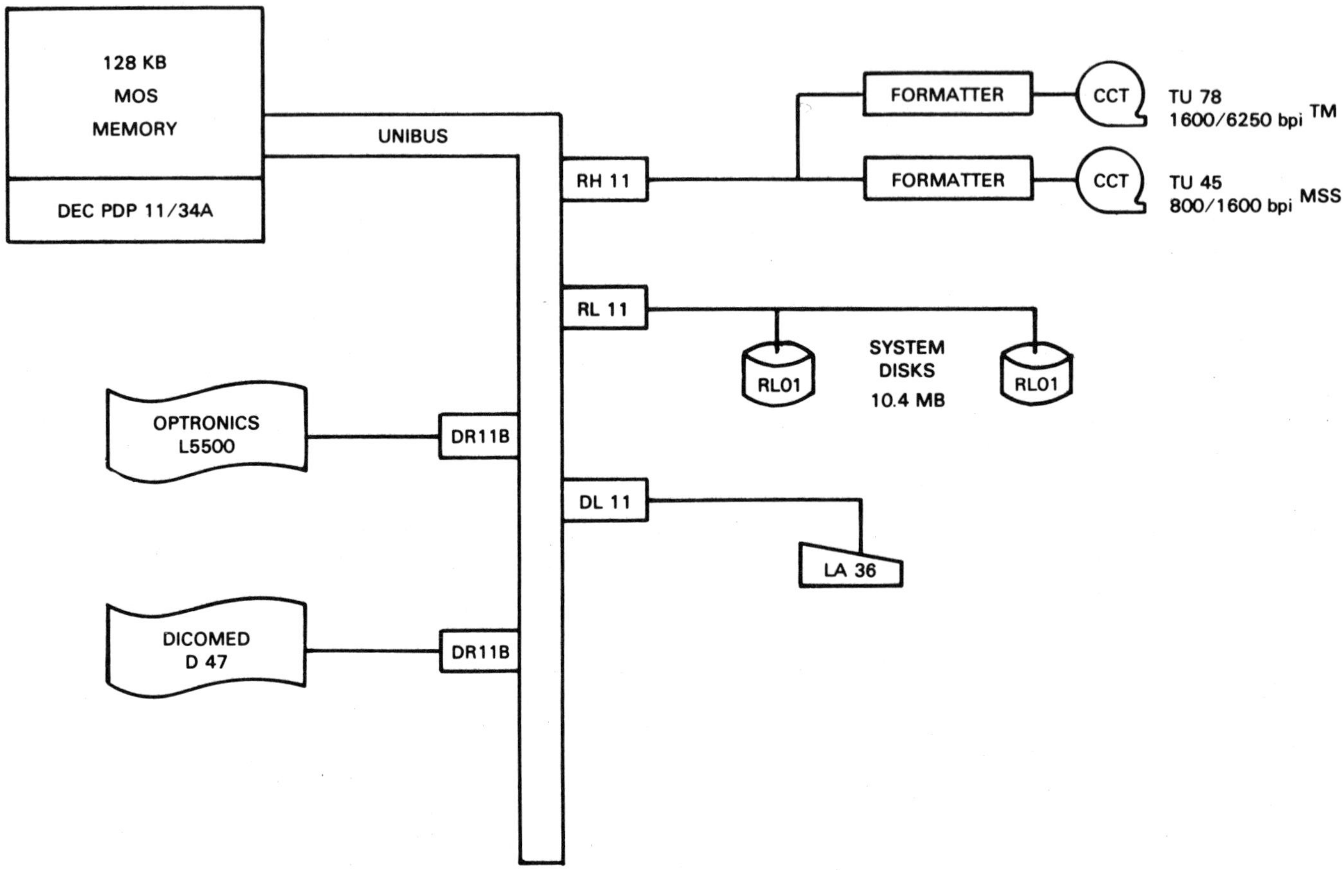


*Figure 4. LAS Film Recorder Subsystem*

tensive file management capabilities. The LAS Executive and the applications support software packages are being designed and developed by Goddard personnel. These software packages will operate in conjunction with the VMS operating system and will make use of VMS capabilities wherever possible.

The LAS Executive will provide the interface between the user and the system and will manage all batch and interactive processing operations. The basic design philosophy of the LAS Executive includes adapting and extending the Video Image Communication and Retrieval (VICAR) system, developed at NASA's Jet Propulsion Laboratory, for use in the LAS. The VICAR extensions currently planned for the LAS will be implemented in FORTRAN IV PLUS and include integrating a menu-driven executive, developing a procedure capability for easy execution of batch or interactive jobs with multiple processing steps, and adding data management capabilities which make it easier to catalog, store, retrieve, combine, display, and manipulate data-bases containing both image and non-image data. The system will provide these capabilities while maintaining the VICAR command language structure and applications program linkage conventions intact. Therefore, to an LAS user, the system will offer an easily used menu-driven or VICAR command language interface with extensive procedure and data management capabilities in support of both batch and interactive processing operations.

This approach offers several major advantages including: reduced LAS software development costs, since all existing VICAR applications programs can be installed in the LAS with minor modifications; the ability to share software with other VICAR users; improved VICAR capabilities and improved user interface with VICAR and standardizing software, software interfaces, and data formats for the LAS and future systems to be developed by our group.

The LAS Executive and planned modifications of VICAR will provide a flexible, easily extended system which offers the basic image-processing and related capabilities needed to support the LAS investigator teams.

New applications software required by the investigator teams will be developed as applications programs under the LAS VICAR system. Where appropriate, versions of LAS processing software will be implemented to utilize the AP-120B pipelined processor as well as the VAX mini-computer.

The specific tasks envisioned for the AP-120B include: geometric correction/image resampling, data set registration/image remapping, principal component analysis/feature selection, image enhancement, multispectral classification, clustering and executing applications models.

The LAS software system will be modular and extendable and should be useful for a wide range of remote-sensing and image-processing applications.

## LAS APPLICATIONS

Applications investigators will use the LAS interactive image display and analysis consoles to access, display, extract, analyze, and evaluate information derived from the LANDSAT-D data and required ancillary data.

Examples of investigations to be conducted on the LAS include demonstrating and evaluating the utility of the TM versus MSS data for the following applications:

*Agriculture*
- crop indentification/classification
- crop acreage estimation
- monitoring crop stress and condition
- yield modeling
- small field crop analysis

*Water Resources*
- delineating hydrological land use classes for watershed modeling studies
- investigating water turbidity and water quality
- surface water mapping
- mapping inland wetlands
- mapping flood-prone areas
- snow and ice mapping

*Geology*
- mineral and petroleum exploration
- geobotanical mapping
- geothermal exploration
- geological structure mapping

*Land Use*
- mapping and inventorying urban land cover
- monitoring and mapping surface mining activities
- general land use mapping

*Forestry and Rangeland*
- inventorying and evaluating forest/rangeland resources
- mapping vegetation cover types
- estimating timber volume
- biomass estimation
- detecting and monitoring insect and disease damage
- monitoring vegetation stress conditions
- acreage estimation

Simulated Thematic Mapper data sets derived from aircraft-borne scanners have been used by a number of researchers to determine the potential of the LANDSAT-D Thematic Mapper to provide more accurate and detailed information for Earth resources investigations. For small fields of one to four hectares which are typical of agricultural field sizes in the Eastern United States, Western Europe, India and parts of Asia, Morgenstern et al[4] show that area measurements derived from simulated TM data with 30 meter resolution are 10 to 13 percent more accurate than those derived from 50 meter data. They also describe results on crop classification accuracy versus spatial resolution and radiometric sensitivity.

Tucker[7] investigated the use of the TM spectral channels for monitoring vegetation at different growth stages and concluded that the TM data will significantly advance the state-of-the-art in satellite remote sensing of vegetation. Salomonson and Hall[5] discussed the improvements to be derived from the use of TM data for hydrological applications.

## PREPARING FOR LANDSAT-D

The enormous task of handling data transmitted at 100 megabits per second and containing 250 million bytes of data for each full TM scene requires sophisticated processing systems which include pipelined processors, large disk storage capacity, and the use of efficient storage media such as optical disks or high-density tape recorders. Fortunately, most users of the LANDSAT-D data will require small segments of images for their applications rather than full TM scenes. For these users, shared use of mini-computers or of large-scale computers will continue to provide the needed capabilities to analyze TM data, as is currently the case for MSS data from LANDSATS 1, 2, and 3.

In order to adapt existing LANDSAT MSS processing software for use with TM data, users will have to perform at least the following modifications:

- Develop new data ingest programs which read the EROS Data Center LANDSAT-D computer-compatible tape format.
- Modify tape reading programs to handle full scenes which span multiple data tapes (a full TM scene will reside on from two to seven computer compatible tapes, depending on the tape density and blocking factors used).
- Develop efficient programs to select, transfer, and display averaged or subsampled overviews of full TM scenes for synoptic viewing and subscene selection.
- Develop efficient programs for reading and storing designated subscenes from full TM scenes to reduce data transfer and storage volume requirements when analyzing image segments.

With these modifications, existing LANDSAT processing systems should be able to ingest segments of full TM scenes and process the segments as is done with current MSS data. Other possible software modifications include increasing data array and program loop sizes to handle up to seven TM spectral channels.

Users should be aware that the VICAR software system provides most of the capabilities needed to process, correct, enhance, and analyze LANDSAT data. The VICAR system has been widely disseminated in the public domain and now includes both an IBM 360/370 version and a PDP-11 mini-computer version developed by NASA's Jet Propulsion Laboratory. Both versions of the VICAR software have been made available at low cost through NASA's COSMIC software dissemination facility.

Software for the modifications required to process LANDSAT-D data under both the IBM 360/370 and the LAS mini-computer version of VICAR will be developed at Goddard and will be available from: COSMIC, University of Georgia, Athens, GA 30601.

## CONCLUSION

The LANDSAT-D system, and particularly the Thematic Mapper, promises to provide greater responsiveness, higher resolution, and improved capabilities for applying remote-sensing data to the solution of problems in monitoring and managing the Earth's resources. One of the largest challenges to be faced in demonstrating the system's capabilities will be handling the huge volume of data associated with the LANDSAT-D 100 megabit per second data rate. The LANDSAT-D ground processing system and the LANDSAT-D Assessment System are being designed and organized to meet the challenges presented by the Thematic Mapper and to meet the need for investigators to interact rapidly with the LANDSAT-D data.

The LANDSAT-D Assessment System will provide a test bed for capable evaluation of the LANDSAT-D data and for developing and demonstrating improved information handling capabilities for Earth resources management.

## ACKNOWLEDGEMENTS

The authors wish to thank V. Salomonson, LANDSAT-D project scientist, D. Williams, assistant project scientist, L. Smith and W. Webb, LANDSAT-D project, J. Billingsley, NASA/GSFC, D. Smith and J. Brooks, the General Electric Company, and the other members of the NASA LAS development team for valuable discussions during the preparation of this paper.

## REFERENCES

[1]Blanchard, E. E., and O. Weinstein. "Thematic Mapper Development." *Proceedings of the 1979 Symposium on Machine Processing of Remotely Sensed Data,* Purdue University, West Lafayette, IN, June 1979.

[2]Bracken, P.A.; J. T. Dalton; J. J. Quann; and J. B. Billingsley. "AOIPS — An Interactive Image Processing System." *Proceedings of the 1978 National Computer Conference,* Anaheim, CA, June 1978.

[3]Fuchs, A. J., and R. S. Pajerski. "The Role of Autonomous Satellite Navigation in the NEEDS Program." *Proceedings of the AIAA/NASA Conference on Smart Sensors,* Hampton, VA, Nov. 14-16, American Institute of Aeronautics and Astronautics, NY, 1978.

[4]Morgenstern, J. P.; R. F. Nalepka; E. R. Kent; and J. D. Erickson. "Investigation of LANDSAT Follow-On Thematic Mapper Spatial Radiometric and Spectral Resolution." Environment Research Institute of Michigan Technical Report 119300-10-F, Ann Arbor, MI, June 1976.

[5]Salomonson, V. V., and D. K. Hall. "A Review of LANDSAT-D and Other Advanced Systems Relative to Improving the Utility of Space Data in Water-Resources Management." Final Workshop on the Operational Applications of Satellite Snowcover Observations, Sparks, NV, Apr. 1979.

[6]Smith, P. L., and W. C. Webb. "LANDSAT-D Acquisition and Data Processing System." *Proceedings of the 1979 Symposium on Machine Processing of Remotely Sensed Data,* Purdue University, West Lafayette, IN, June 1979.

[7]Tucker, C. J. "A Comparison of Satellite Sensor Bands for Vegetation Monitoring." *Photogrammetric Engineering and Remote Sensing* 44, pp. 1369-1380, 1978.

# A Case Study of an Application of Computer Mapping in Oil-Shale Resource Mapping

*by Frank G. F. Davis, Jr. and John Ward Smith*

## ABSTRACT

The Laramie Energy Technology Center, U.S. Department of Energy, is responsible for evaluating the resources of potential oil and the deposit characteristics of oil-shales of the Green River Formation in Colorado, Utah, and Wyoming. While the total oil-shale resource represents perhaps two trillion barrels of oil, only parts of this total are suitable for any particular development process. To evaluate the resource according to deposit characteristics, a computer system for making resource calculations and geological maps has been established. The system generates resource tables where the calculations have been performed over user-defined geological intervals. The system also has the capability of making area calculations and generating resource maps of geological quality. The graphics package that generates the maps uses corehole assay data and digitized map data. The generated maps may include the following features: selected drainages, towns, political boundaries, township and section surveys, and corehole locations. The maps are then generated according to user-defined scales.

## BACKGROUND DESCRIPTION AND DEFINITION OF PROBLEM

Oil-shale is a source of liquid oil refinable into petroleum products. The United States has a tremendous resource of oil-shale, estimated to represent 3 x $10^{12}$ barrels of potential oil in place in oil-shale rocks. About 20% of the United States is underlain by deposits described as oil-shale, but the largest and richest deposit in the world is the Eocene Green River Formation. This 16,500 square mile deposit in adjoining corners of Colorado, Utah, and Wyoming represents two thirds of the United States oil-shale resource. However, these resource values are gross estimates. Development of oil-shale to produce oil requires much more precise knowledge of the resource. The Laramie Energy Technology Center (LETC), U.S. Department of Energy, has the responsibility of defining this resource and its developmental potential. This responsibility is being met with the aid of the computer data processing and graphics described.

The Green River Formation deposits are characterized by lateral homogeneity, exhibiting only gradual geographic changes in organic and mineral composition. Vertically, however, the formation is extremely variable, and this vertical variability influences the applicability of development methods. These characteristics, together with the large area covered, present the problem of evaluating the resource efficiently. This must be accomplished rapidly and flexibly in order to be useful in planning production experiments and generating energy policy leading to commercial development of the resource.[6]

The basic tool for oil-shale deposit evaluation is oil yield, determined by an analytical process known as Fischer assay.[5] Oil-shale has no oil in it. Instead it contains an insoluble solid organic material which breaks down to yield oil when the rock is heated. The Fischer assay procedure developed and standardized by LETC is an empirical method of estimating the amount of oil which might be produced from a sample by some commercial process. Over 35 years of continuous oil-shale assaying, LETC has developed an extensive bank of oil yield assay data. This information, some of it published,[2 3 4 6 7] provides the basis for resource and deposit evaluation. The last reference refers to additional available data.

Oil yield data from Fischer assay is reported as gallons of oil per ton of shale. This weight unit is useful analytically, but the oil-shale deposit must be evaluated by volume instead of by weight. Because of the variable content of organic matter in oil-shale, the density of the oil-shale rock varies from 1.5 to 2.7. An equation relating rock density to oil yield from Fischer assay provides the density estimate necessary for deposit and resource evaluation.[1]

Due to the magnitude of the problem in terms of both area and number of data points, the computer application was initially restricted to the Piceance Creek Basin of Colorado. This area is approximately 1340 square miles. The Basin contains the richest deposits of Green River Formation oil-shale, up to 2100 feet (700 meters) thick, with oil yields varying from a few gallons of oil per ton up to almost 100 gallons per ton. The deposit reaches depths of 3500 feet. By 1976 assays from about 230 coreholes representing over a million pieces of data were available.

The problem stimulating this progress report is how to make

resource evaluations and to generate resource maps of geological quality. These maps must be highly accurate in terms of user-specified geological features, resource information, and well locations. Additionally, the maps need to include such items as land survey data, drainages, towns, and political boundaries.

## METHOD DESCRIPTION

An evaluation of a resource is accomplished by obtaining continuous cores from wells drilled at selected sites, assaying the cores at variable intervals averaging about one foot, measuring the area covered, and then calculating the amount of oil in the area that would be available.

An essential part of the process of evaluating the Green River Formation oil-shale deposits is exploiting the lateral homogeneity and the vertical variability of the oil-shale.[3,6] This requires generation of histograms of oil yield vs. depth for each well within the selected area. Geological markers are then picked and correlated from well to well across the deposit. Once the geological markers have been correlated, calculations are made to determine the oil yields over corresponding intervals of core within a well. From this information, oil yield histograms are prepared. When done by hand a histogram requires several days to prepare. The long time results from the need for accuracy in both the drafting and the proofreading. The evaluation of the Piceance Creek Basin Green River Formation involved over 200 wells; this would have required a long preparation time and the handling of large amounts of data. The process was speeded up by preparing a plotting program to generate the oil yield histograms of individual wells (Figure 1). These plots are then used as the basis for the deposit correlation (Figures 2 and 6).

Once the correlation is completed, it is possible to specify the deposit characteristics of the area for which the resource evaluation was being prepared. From this it becomes possible to generate resource maps of geological quality if the necessary computer equipment is available. An outline specifying the requirements that a mapping system would have to meet was prepared. Computer Research Corporation (CRC) was selected to provide the specialized software and hardware. They began by constructing a data-base of different types of data files. First, township-section survey data were digitized from U.S. Geological Survey (USGS) 7½ minute topographic maps. Then the political or feature boundaries of the area were digitized (Figure 3). Separate files for each special feature of interest were prepared. Finally, the survey data for the location of each well was converted to latitude-longitude decimal coordinates. Then by a process of computer overlays a map of the area of interest or any desired section of interest may be created (Figures 4 and 5).

A map resulting from the above process can be scaled according to use specifications. It can be combined with the data resulting from the oil-yield calculations and have computer contours generated across the maps (Figure 7). A separate contour map of the calculated yield of oil can be prepared for each geological marker. It then becomes possible to calculate the areas covered by a particular yield. This information is then used to calculate the total resource available under any kind of geological or technological specification.

## RESOURCE REQUIRED

Primarily the resources of three institutions have been involved. The Laramie Energy Technology Center (LETC) for specifications of the system, funding for the entire project, acquisition of the data, assays of the data, storage of data on the computer, six 25 megabyte disc packs for data storage, and scratch tapes. The University of Wyoming for a Xerox Sigma 7 Computer, Calcomp Drum Plotter, and Versatec Plotter. The Computer Research Corporation for a graphics package, digitizer, PDP computer, and flatbed plotter.

## PROBLEMS ENCOUNTERED

The problems encountered were: communicating to potential vendors what our needs were, data problems, conversion from an old mainframe computer to a new one, and modification of a contractor program for a specific requirement.

The vendor communications problem was solved by presenting each potential vendor with an outline of our needs, both present and future. This was done at LETC so that a question and answer session would develop between the vendors and the potential users. It also allowed the vendors to show examples of the capabilities of their systems.

A variety of data problems occurred. They were largely due to the lack of standarized format; data file names that could not be traced to the laboratory assay tables, resulting in duplication of the data files; accuracy of data input into the computer; and surveyed well locations that were expressed in any one of six different methods: (1) latitude-longitude degree coordinates, (2) feet and degrees from a section corner, (3) feet from a section corner (not necessarily the one in which it is located), (4) feet from two sides of a section, (5) the specification of just the section in which the well is located, and (6) the specification of quarters of a section in which the well is located. The solutions resulted in the reworking of every data file in the database. The process began by defining a standardized data format. In addition to the assay data the format included the information records at the beginning of a file containing descriptive information such as file name and well location. The file names were modified to include a code for the state, a code that indicates the core assay data, and a unique number assigned by the assay laboratory that generated the assay data. The survey data was included so it could be converted by the computer to latitude-longitude decimal coordinates.

The original plans called for installing the contractor's programs on a Xerox Sigma 7 Computer. Maps would then be interactively created through the use of a graphics CRT. A tape of the map codes would then be sent to the contractor to be output on a flatbed plotter. A problem arose when the University of Wyoming decided to replace the Xerox Sigma 7 with a system of two CDC computers. Rather than go through the problems and expense of installing the contractor's programs on the Sigma and later a CDC, it was decided to wait for the new machine to be installed. This alternative was chosen because the contractor was willing to generate the maps in the interim.

The final problem arose when a need developed for the capability of indicating with dashed lines whenever a township or section line on the USGS maps was missing. This requires a software modification still being worked on.

## DOCUMENTATION

Parts of the documentation are available from two sources: assay data formats, LETC program documentation, and research reports are available from LETC. Mapping software documentation is available from DRC.

## RESULTS

Formerly oil-shale resource calculations and the generation of associated geological maps were a major roadblock to the development and publication of a resource evaluation. The process would usually take years. The computer has removed the roadblock of time. Including cross checks for accuracy, the process now takes less than a month.

There has also been a tremendous increase in the accuracy and validity of the results. So much so that a detailed resource analysis can now be incorporated in the planning stages of retorting experiments and energy policy.

## MANAGEMENT AND STAFF REACTIONS

Management was initially faced with a number of problems: large amounts of data, the necessity of long lead times (years) to produce resource evaluations and geological maps, tighter budgets, and the use of large numbers of man-hours. It was felt that there were no individual solutions to the above problems in the foreseeable future. As a result, management made the decision to investigate the use of the computer as a solution to the problems. Once the decision had been made, management backed it and delegated the necessary responsibility. The reaction when the first results began to appear was, "It is just what was needed. Now, can you do...?"

The reaction of the staff has been positive. It helps to keep the motivation up for a project when the staff can receive rapid feedback.

## HINDSIGHT

Two lessons were learned that would be of use in any computer project. First, a large amount of time must be committed to the quality of the initial data formats and the validity of the data. All subsequent work will depend on the flexibility of the formats and the quality of the data. Second, the specification of the needs and the evaluation criteria that the project will be judged by should be as clear and specific as possible. This will avoid a lot of unnecessary confusion and discussion between the users and potential vendors.

## EXPERIENCE AND THE FUTURE

The experience gained on this project will be of use in the future. This project will be repeated while being extended. Work is beginning on duplicating it for the Green River Formation oil-shale areas of Utah and Wyoming.

The procedures used here should be of use to anybody planning to conduct a resource evaluation that has a similar data requirement. As an example, they are expected to be used by the tar sands and coal gasification programs that are being developed at LETC.

## REFERENCES


[1]Smith, John Ward. "Specific Gravity Oil Yield Relationships of Two Colorado Oil-shale Cores." *Ind. Eng. Chem.*, Vol. 48, 1956, pp. 441-444.

[2]Smith, John Ward, and Howard B. Jensen. "Oil Shale." *McGraw-Hill Yearbook of Science and Technology*, 1977.

[3]Smith, John Ward, T. N. Beard, and L. G. Trudell. "Colorado's Primary Oil-Shale Resource for Vertical Modified In-Situ Processes." *National Technical Information Service*, LERC/RI-78/2, 1978.

[4]Stanfield, K. E., J. W. Smith, and L. G. Trudell. "Oil Sections of Green River Oil Shale in Colorado, 1957-1963." U.S. Bureau of Mines. RI-7051, 1967.

[5]Stanfield, K. E., and I. C. Frost. "Method of Assaying Oil Shale by a Modified Fischer Report." U.S. Bureau of Mines, RI-4477, 1949.

[6]Trudell, Laurence G., Thomas N. Beard, and John Ward Smith. "Green River Formation Lithology and Oil-Shale Correlations in the Piceance Creek Basin, Colorado." U.S Bureau of Mines, RI-7357, 1970.

[7]*A Bibliography of Publications Dealing with Oil Shale and Shale Oil.* U.S. Bureau of Mines, 1917-1974, and the ERDA Laramie Energy Research Center, 1975-1976. LERC/RI-77/7, 1977.

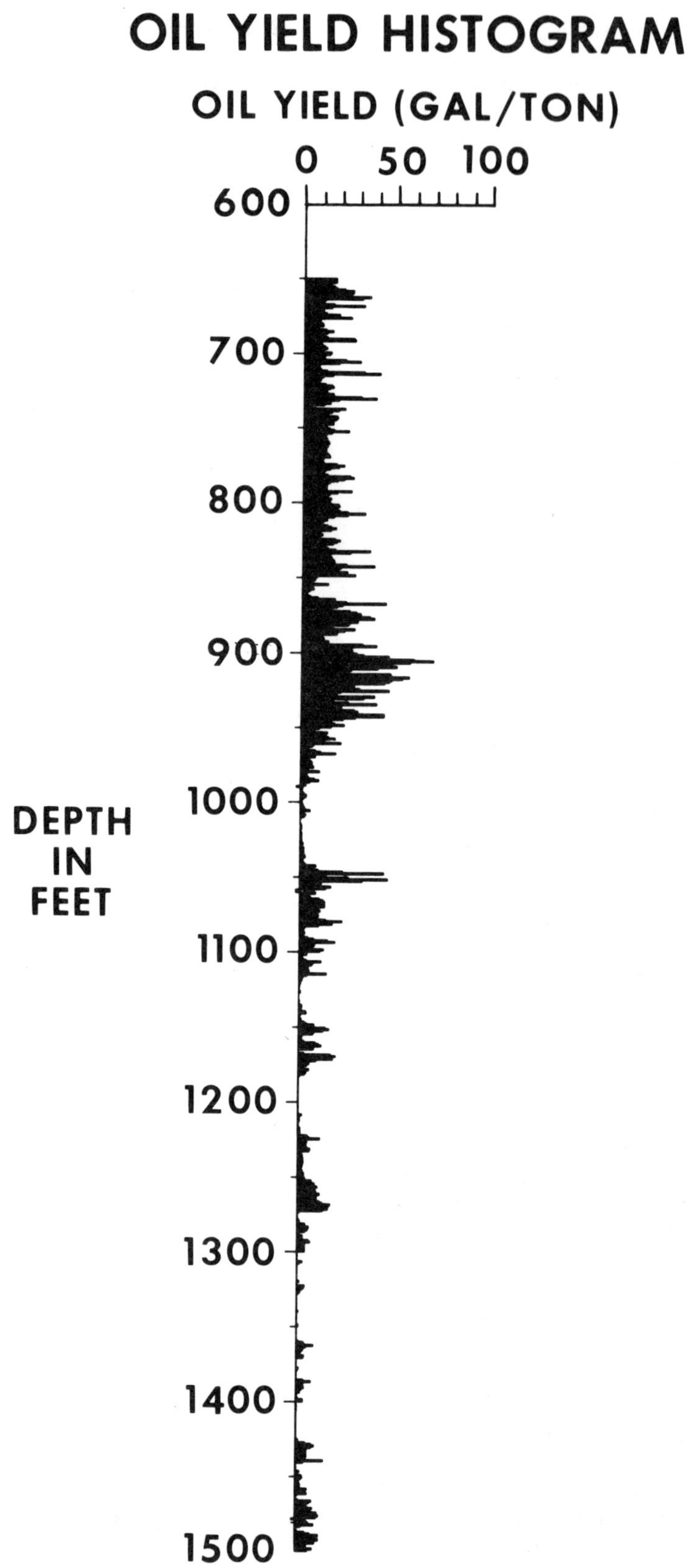


*Figure 1. A Histogram of the Fischer Assay Oil Yields for a Core Well*

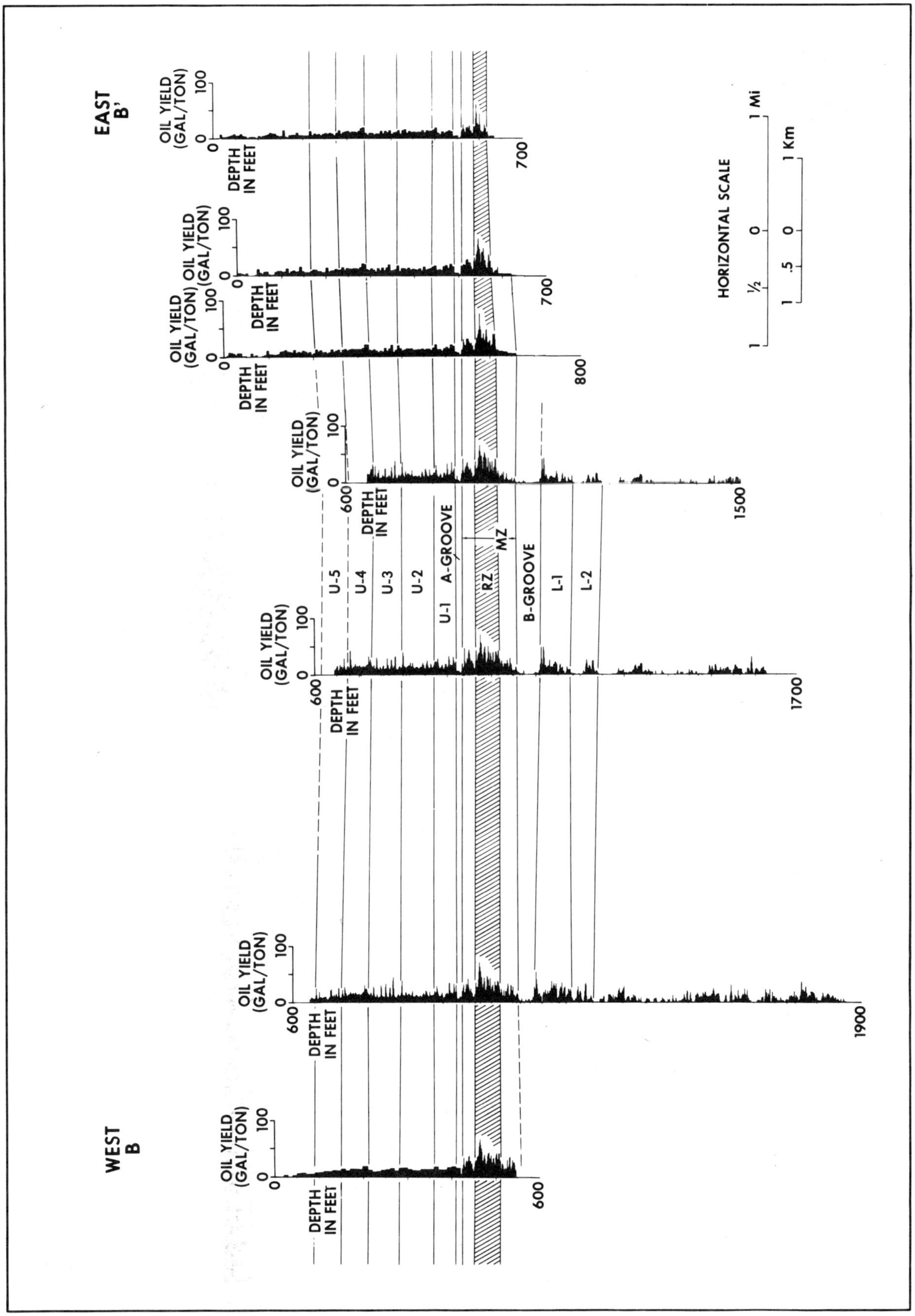


*Figure 2. Correlations between the Histograms of Core Wells*

*Figure 3. Mahogany Marker Outcrop, Piceance Creek Basin, Colorado*

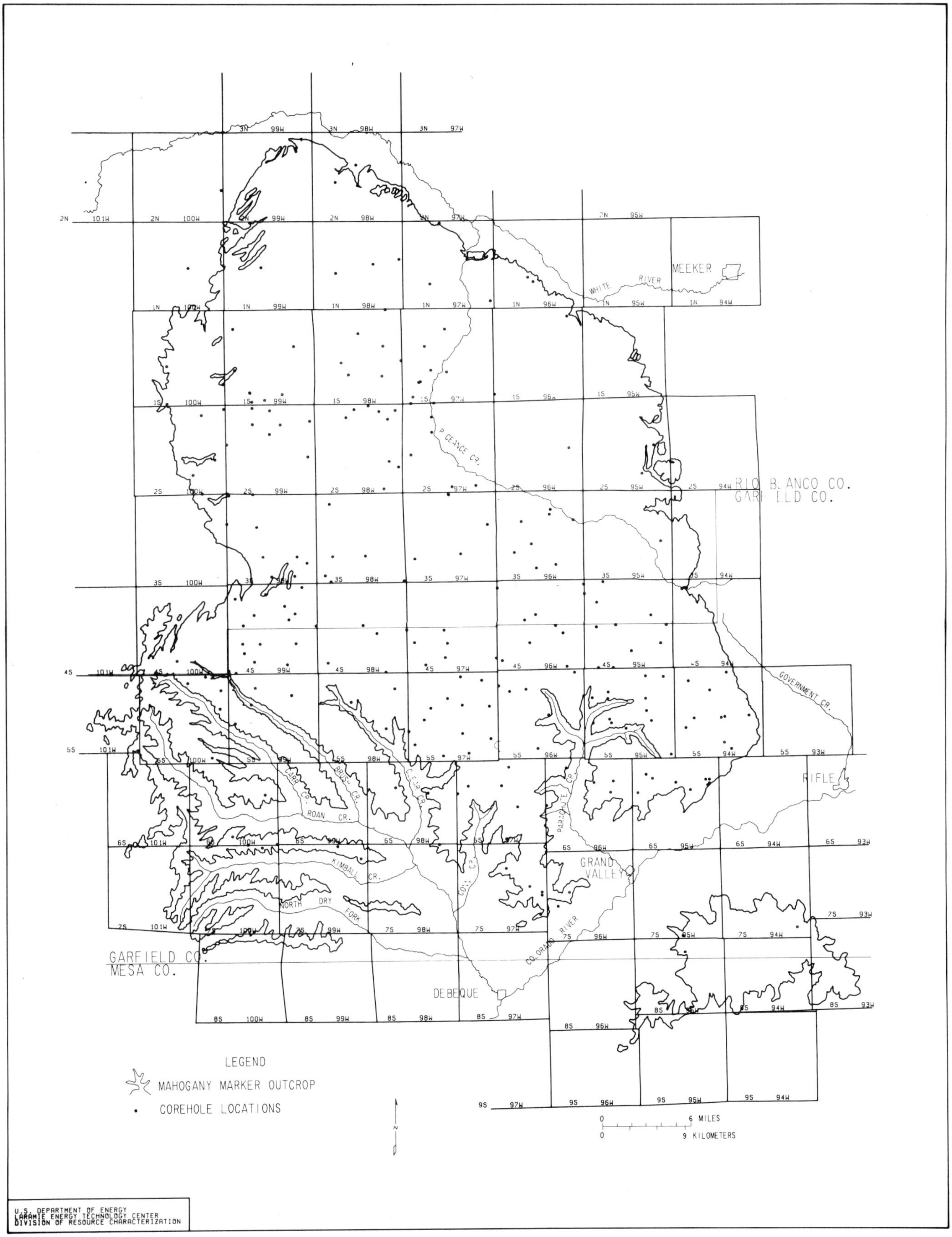


*Figure 4. The Mahogany Marker Outcrop Map with Townships, Well Locations, and Additional Details Added*

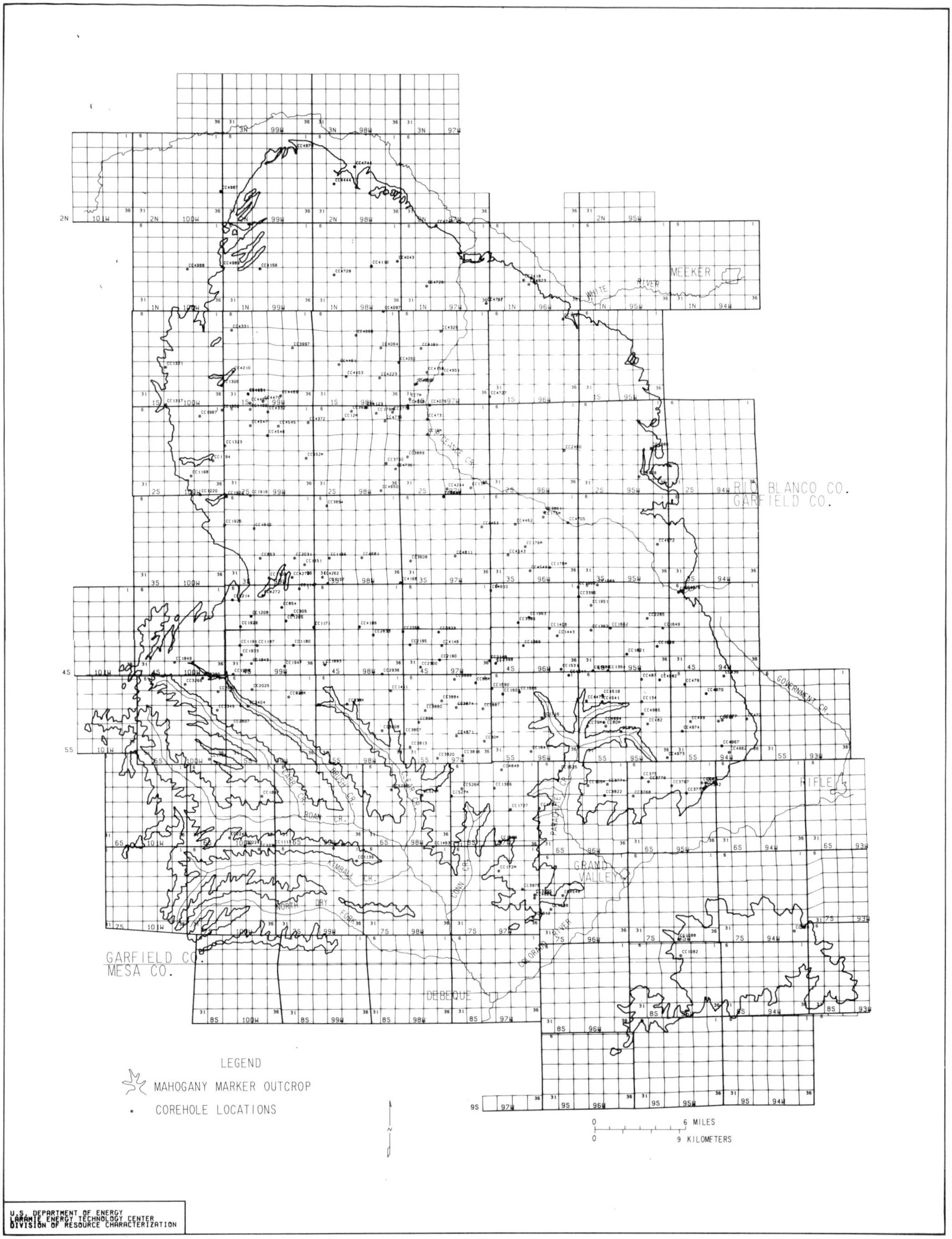


*Figure 5. The Mahogany Marker Outcrop Map with Sections and Assay Data File Names Added*

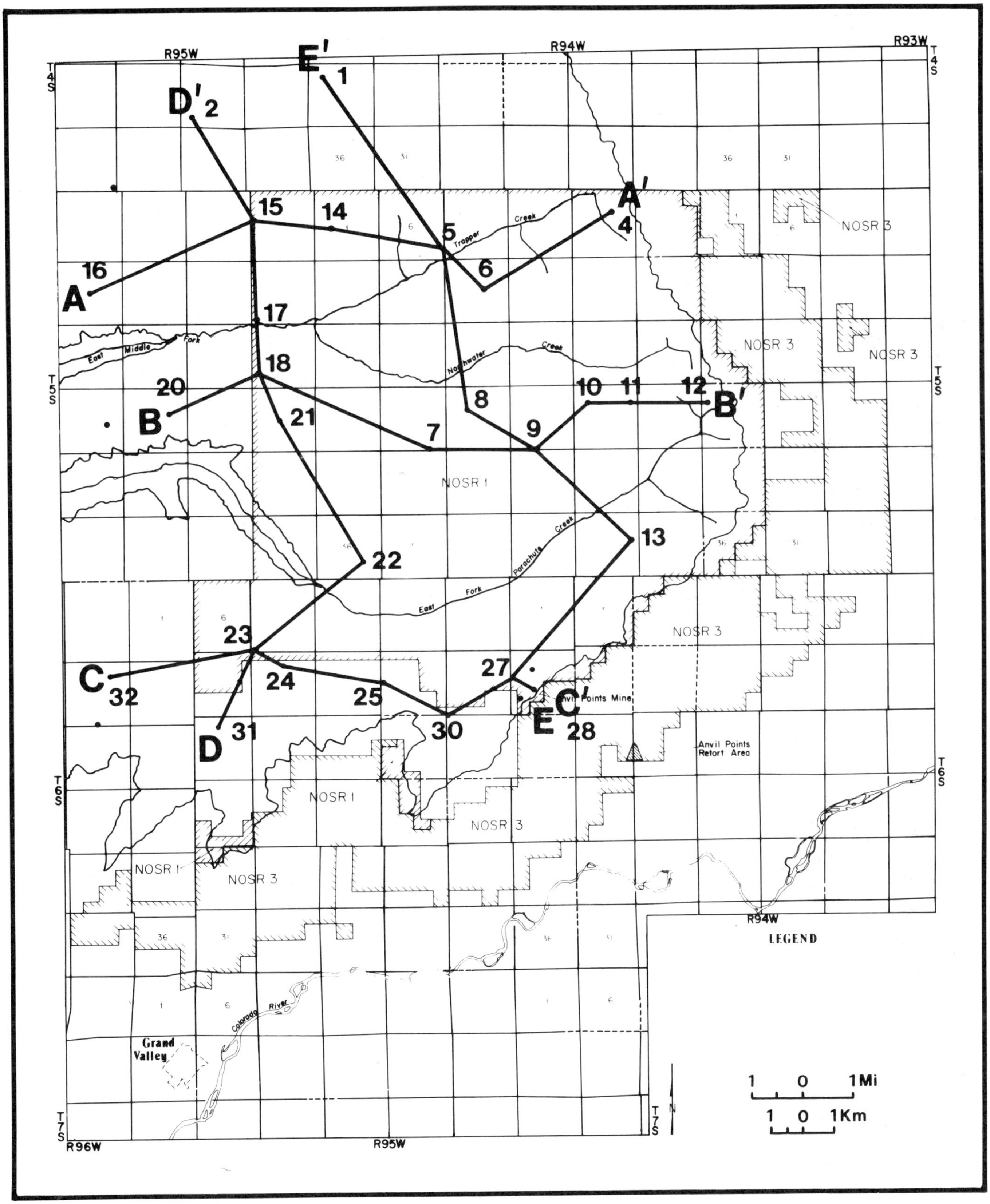


Figure 6. Naval Oil Shale Reserve No. 1, a Portion of the Piceance Creek Basin, with Correlated Wells Indicated

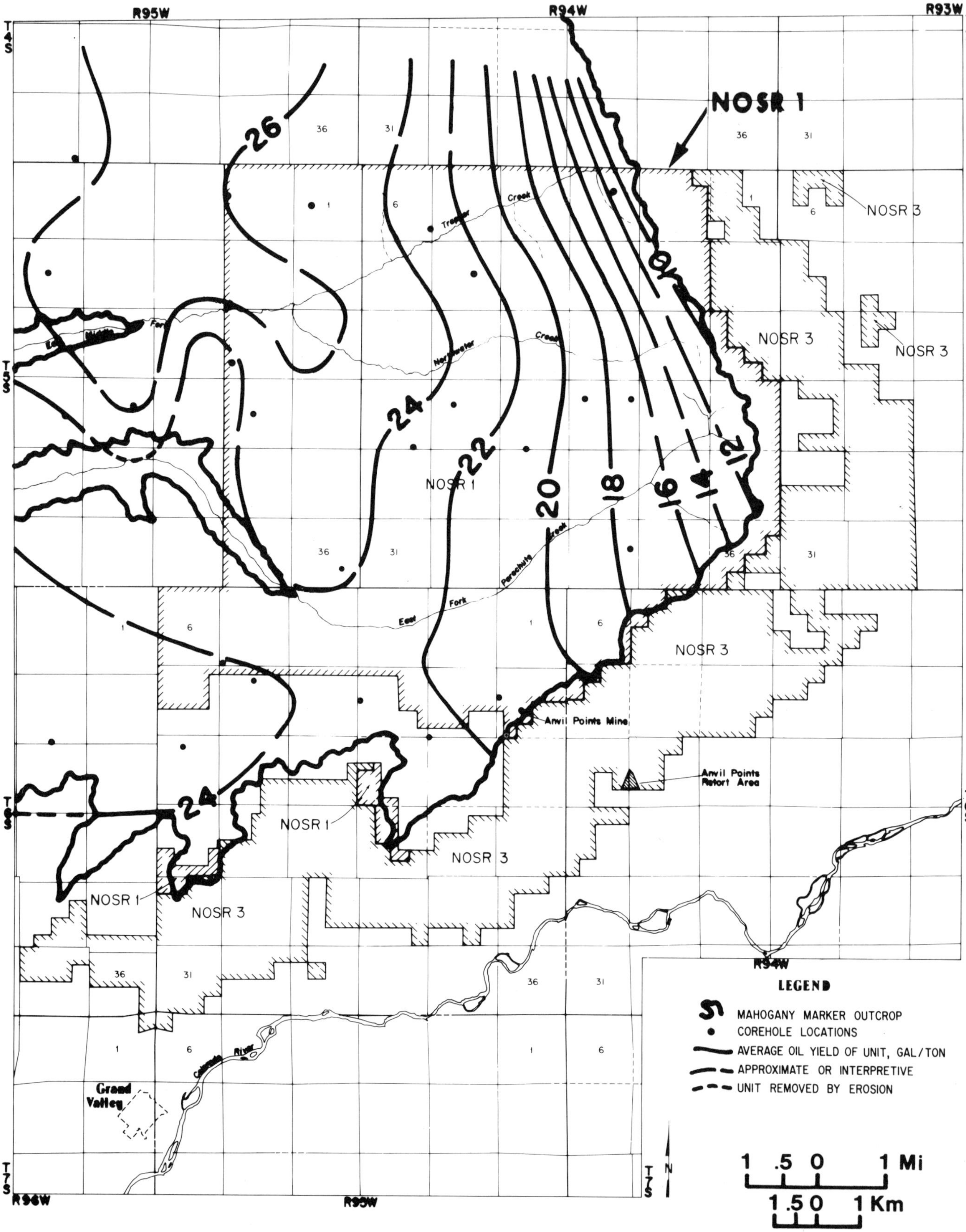


*Figure 7. Naval Oil Shale Reserve No. 1, with Oil Yield Contours*

# Remote Sensor Inputs to Information Systems Research at the National Aeronautics and Space Administration

*by John E. Estes and Ruth I. Whitman*

## INTRODUCTION

At the time of this symposium, the field of remote sensing has been in existence for some sixteen years. Yet to date there has been only limited theoretical discussion of remote sensing as a source of data for geographic information systems. Even less discussion has occurred concerning the techniques and advantages of direct linkage between geographic information systems and remote-sensing information acquisition processing and analysis systems. Yet geographic information systems are becoming important tools in public decision-making and have significant potential to incorporate and facilitate the use of remotely sensed data. This potential has only barely begun to be explored, let alone developed. For example, recent developments in satellite remote sensing have established pixels as locational identifiers, produced large volumes of data in digital format, and created new types and frequencies of data for environmental anlaysis. These data could and should, routinely and directly, be used as input to geographic information systems. However, to date there has been conspicuous lack of understanding of a number of important factors, such as the data input problems of the geographic information systems; the monitoring capabilities of remote-sensing systems; and the potential for improved modeling and analysis of environmental conditions and processes, based on greater utilization of the monitoring capabilities of remote sensing and the analytical capabilities of computer-based information systems (Shelton and Estes, in press). This paper details the role and present activities of the National Aeronautics and Space Administration (NASA) in information systems research in the land resources area.

## NASA'S ROLE IN LAND RESOURCES

The inventory, analysis, and assessment of the level of human use of land encompasses the application of information from many disciplines, including agriculture, hydrology, geography, forestry, rangeland management, and geology. Systems for environmental monitoring attempt to identify the complex relationships of people's activities to their surroundings. Cartographic tools are employed in planning, implementing, and monitoring nearly every human activity involving interaction with the physical environment (National Academy of Sciences/National Research Council, 1976). Today we have the remote-sensing capability to map land use/land cover categories effectively for the entire United States, at regional scales within a reasonable time frame compatible with the needs of land resources planning, management, regulation and research personnel (Anderson, 1976). Remote sensor systems from the LANDSAT series, high altitude platforms, and other sources, employed together with collateral data, represent a land resources data collection system capable of providing cost effective data on current and changing land use/land cover patterns, which can be structured in georeferenced information systems so that the spatial and functional dynamics and trends in land resources parameters can be efficiently studied.

One of NASA's primary roles in this process is to augment the existing land resources analysis and management process by promoting the incorporation of space-acquired data. Owing to the diversity and complex interrelationships of land resources applications, the accomplishment of NASA's goals is being pursued through a basic SR&T program directed at major research areas including: land use/land cover classification; the improved input of remotely sensed data into geobased information systems; land use/land cover change detection; and land resources modeling.

These generic research areas may be thought of as independent and yet interdependent. In some respects, these areas may be viewed in a somewhat hierarchical manner; e.g., if we improve our capability to accurately classify, in an absolute sense, land use/land cover data, this will lead to an improved capability to input these data into geobased information systems, to detect change, and to derive more reliable information concerning land capabilities and predictive estimates of future changes (Figure 1). Yet, in a very real sense, these research areas are symbiotic and synergistic in nature, easily coexisting, and advances in any area benefit the whole (Figure 2).

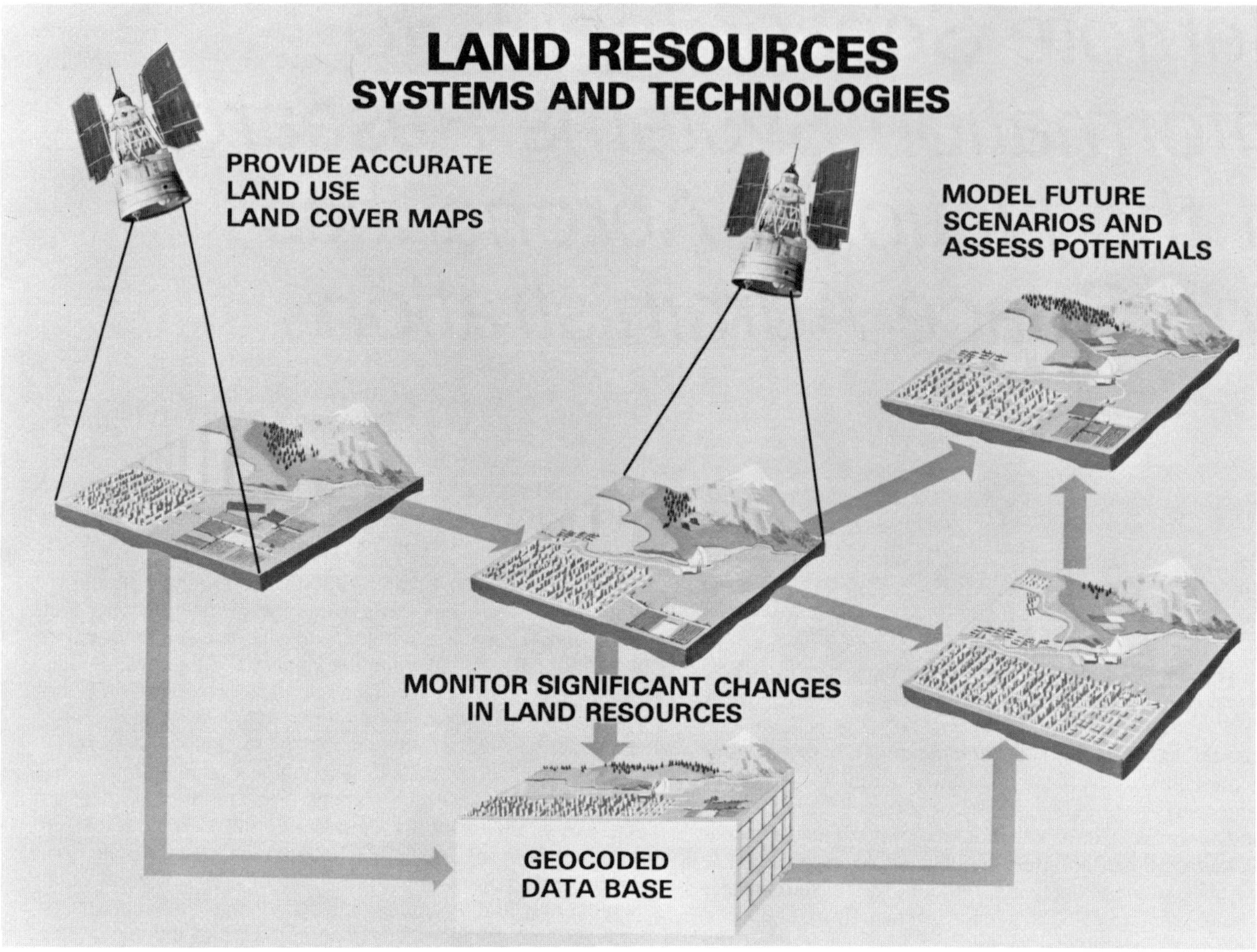


*Figure 1. Land Resources Systems and Technologies Geocoded Data-Base*

Groups that may benefit from remote-sensing research in those areas outlined above include local agencies such as county- or area-wide Councils of Government (COG), planning departments, State Offices of Planning and Research, Departments of Water Resources and Natural Resources, regional entities such as River Basin Commissions, federal agencies such as the Department of the Interior, Department of Commerce, Department of Agriculture, Department of Energy, National Oceanic and Atmospheric Administration, Environmental Protection Agency, Housing and Urban Development, and the Corps of Engineers, as well as public utilities and private consulting firms with socioeconomic, biophysical and engineering interests. In order to develop space remote-sensing data into land resources information, a significant research program will be required. NASA believes this program will be most efficient and effective if a cooperative program is developed between NASA, significant user groups, universities, and private industry.

Although in recent times NASA has not employed the research potential of universities in this area to its optimum extent, NASA supports and encourages the concept of close cooperation, not only with specific potential users of the results of this SR&T program (e.g., Geography Division of the Bureau of the Census and Geography Program of the United States Geological Survey), but also with university and private industrial researchers.

As NASA conducts its research in these areas, the user's desire for accurate land use/land cover maps potentially derived from the remotely sensed data as well as tabular and statistical data must be remembered. We are working to gain a thorough understanding of the factors affecting the cartographic accuracy with which thematic data derived from remotely sensed data may be presented. This work is extremely important as we attempt to more accurately input remotely sensed data into information systems. At present, geometric analysis of LANDSAT RBV and MSS imagery indicates that a capability of 1:250,000-scale monochromatic image-mapping at National Map Accuracy Standards[1] under proper conditions exists (Colvocoresses, 1976). In the future,

[1]Horizontal accuracy: For maps at scales larger than 1:20,000, 90% of all well-defined features will be located within 1/30 inch (0.85mm) of their geographic locations as referred to the map projection; for maps of smaller scales, 1/50 inch (0.5mm).

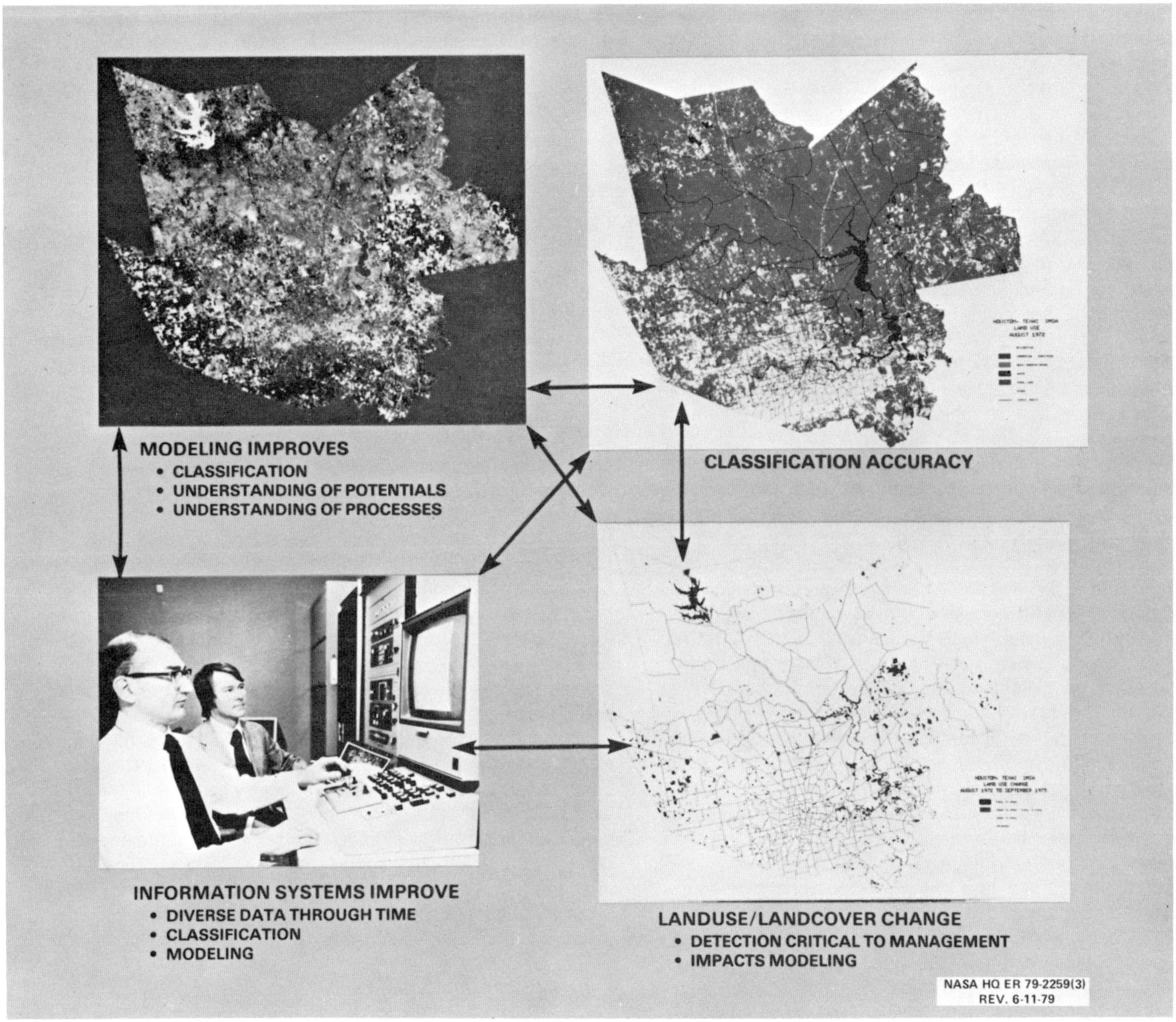


*Figure 2. Land Resources Program Interfaces*

as NASA orbits higher resolution systems capable of greater geometric fidelity (e.g., Thematic Mapper and the proposed Multispectral Resource Sampler), we should see thematic maps produced which meet National Map Accuracy Standards at 1:100,000 and 1:63,360.

Again, as NASA strives toward these goals, the accuracies with which we may interface remotely sensed data with other information in georeferenced data-bases will improve. It is significant to note here that NASA's major role in this SR&T program, with respect to georeferenced or geobased information systems, is not that of improving the structure or conceptual framework of these information systems, but is specifically directed toward improving our techniques for inputting classified remotely sensed data into such systems and achieving more effective integration of these data with data of other types from other sources (e.g., digital terrain data and census information).

Vertical accuracy: 90% of all contours and elevations interpolated from contours accurate within one-half of the basic contour intervals. Discrepancies in these accuracies beyond this tolerance may be decreased by assuming a horizontal displacement within 1/50 inch (0.5mm).

## *IMPROVED REMOTE-SENSING INPUTS TO INFORMATION SYSTEMS*

During the past fifteen years a growing number of federal, state, and local planning and resource agencies have adopted or shown interest in the use of Geobased or Georeferenced Information Systems (GIS) to assist them in making effective decisions concerning land use/land capabilitiy issues. GIS may be of a manual, computer-aided, or interactive computer-aided mode, or a combination of these, and can accept tabular and/or graphic (including remote sensor) data as inputs. Like most management information systems, a GIS permits the user to acquire, encode (format), store, manipulate, and display data. However, geobased information systems are unique in that data are referenced to a common geographic base (e.g., Universal Transverse Mercator (UTM), state plane coordinates, and/or longitude/latitude) which permits storage and retrieval of all data by spatial location. Recently, computer-aided automatic geographic information systems (AGIS) using a digital computer to input data have been developed. These continue to receive wider acceptance by decision-makers in government and the private sector who must process and access large volumes of land use/land cover information and attempt to analyze complex data interrelationships. Digital and nondigital aircraft and satellite imagery, when integrated with other types of data (e.g., graphic and tabular), has the potential for greatly increasing the efficiency and accuracy of present day geobased information systems. Unfortunately, most image-processing systems on the market today that provide enhancement or classification of images merely produce another image or series of simple tabulations as output. Only Jet Propulsion Laboratory's (JPL) Video Image Communication and Retrieval System (VICAR) and Image Based Information System (IBIS), and Electromagnetic Systems Laboratory's (ESL) Interactive Digital Image Manipulation System (IDIMS) attempt to merge remotely sensed data in their digital form directly into a geographic information system format. These attempts have met with some success, but have not adequately demonstrated research feasibility in this area. The major obstacle to full integration of remote sensor and other data inputs to geobased information systems is that image-processing systems are oriented toward raster data structures while GISs use point, line, and polygon structures. No transfer systems that can accept image-processing outputs and convert and reformat these data for input to a GIS have yet been developed and proven effective in a large-scale test. This problem is being worked on, and researchers at Harvard and at the University of California-Riverside are making significant progress in this area. If this problem can be solved, then large volumes of regularly scanned, digitally formatted LANDSAT data will become accessible to decision-makers.

In addition, the need exists to improve our capability to interface remotely sensed data directly with other digital data files (e.g., terrain, soils, and census data) as well as to more adequately correct multispectral scanner (MSS) and digital Synthetic Aperture Radar (SAR) data for horizontal displacement errors caused by variations in topographic relief.

## *PRESENT CAPABILITIES*

Geobased information systems are presently used for a variety of land capability/land use tasks, including land use inventories, agricultural cropland acreage estimates, land carrying capacity/suitability studies, population estimates, and environmental impact assessments. Representative examples of operational GIS are the U.S. Bureau of the Census DIME system, Oak Ridge Regional Modeling Information System (ORRMIS), the U.S. Geological Survey's GIRAS, the Canadian Geographic Information System (CGIS), New York's Land Use and Natural Resources Inventory (LUNR), the Minnesota Land Management System (MLMIS), San Diego's Land Use Management Information System (CPO), and Los Angeles' Land Use Management Information System (LUMIS). The latter six currently employ aerial photography as inputs to their databases. One of the most ambitious uses of remote sensor data to date has been New York's LUNR system, which provides a computerized record of the state's land resources based on a 1967 aerial survey. The system is supported by retrieval, analysis, and display computer programs which record 134 variables for some 140,000 kilometer square cells throughout the state.

At the present time, NASA is supporting a variety of information systems research, Applications Pilot Tests (APT), Applications Systems Verification Test (ASVT), and Technology Transfer projects. These include the Texas Natural Resource Information System (TNRIS) ASVT being done with Johnson Space Center (JSC); California Integrated Resource Information System (CIRIS) ASVT at Ames Research Center (ARC); Census Applications Pilot Test (APT) at Goddard Space Flight Center (GSFC); Navajo APT at National Space Technology Laboratories (NSTL); and work on improving NASA large area LANDSAT mosaicing capability being conducted at Jet Propulsion Laboratory (JPL). These projects have been transferring proven manual techniques for direct digitization of rectified classified LANDSAT data and examining research topics on the accuracy of registration technique and the development of point to polygon and polygon to point conversion software for efficient manipulation of LANDSAT with other data. They have yet, however, to demonstrate adequately the research feasibility of directly inputting radiometrically corrected, geometrically rectified, classified LANDSAT digital data into an information system in such a way that it may be easily interfaced with and interact with other data in the system. This feasibility, in a research mode for accurate and efficient manipulation of classified LANDSAT data in digital form in an information system, is high on NASA's Land Resources Program priority list. Indeed, several research programs will be initiated in the coming fiscal year (FY 80) directed at positions of this task.

To date, LANDSAT data, which is imaged regularly and provided in a digital format, has not received general acceptance in user communities which require spatially accurate land resources data through time. A major problem experienced thus far in integrating remotely sensed LANDSAT data into geographic information systems has been the nonavailability of common georeference formats. For example, pixels of a digital image must be made compatible with other information input to a data-base before the satellite product can be truly useful. Data volume is also a problem because a single LANDSAT scene contains some seven and one-half million pixels, which may exceed the total capability of a user agency's information system. Other problems with LANDSAT imagery that have restricted its incorporation into automated and manual geobased information systems include data format, resolution, image quality, registration/geometric correction, availability of thematically classified data, classification accuracy, and nonstandardization of categories of classified data. The U.S. Bureau of the Census and NASA are presently conducting a Census Urbanized Area Application Pilot Tests (APT) Project designed to evaluate the utility of LANDSAT digital data for providing rapid, cost effective delineation of urban boundaries.

Several approaches to data formatting and manipulation are being assessed, including the potential applications of combining LANDSAT and ancillary data to enhance output products, provide area statistical summaries, and allow for spatial and statistical analysis within the context of an AGIS (Computer Sciences Corporation, 1978).

## *RESEARCH OBJECTIVES*

In order to demonstrate research feasibility and accomplish a variety of demonstration projects illustrating the potential for direct digital input and efficient interaction with other digital data within a GIS, NASA is at present planning to pursue a variety of research topics. One high priority research area is, as previously mentioned, the development of efficient raster to graphics conversion software, designed to transform thematically classified digital image files to arc segment and double digitized boundary vector files at variable grid sizes. Research must also be directed toward improving procedures for making terrain and other digital environmental and socioeconomic data compatible with NASA spaceborne remote sensor data. Another area of essential research deals with the geometric transformations which should be applied to remotely sensed data to correct for displacement errors caused by topographic relief, thereby improving the geometric fidelity of the data. In the future, it will also be necessary to develop software to convert standard LANDSAT series and future satellite data to formats acceptable to existing information systems, and to explore methods for creating derived information planes from multiple-source remote sensor data.

## *CONCLUSION*

NASA is aware that remote sensing is a somewhat unique technology whose transfer to potential users has been conditioned by the U.S. government's role in developing the technology and making its products available. Integration of remote-sensing data acquisition systems, particularly from satellite sensor systems — with geographic information systems (data-processing systems) — has thus consistently been subject to a variety of institutional and technical constraints. As a number of recent reports suggest, however, the full potential of remote sensing will not be achieved without continued efforts to adapt remote-sensing systems-processing and analysis techniques to the evolving needs of users around the world. To the extent that geographic information system designs reflect those needs, GIS design ought to and will be a relevant concern in development of new satellite systems and in establishment of institutional arrangements for processing, formatting, and disseminating the products of remote sensing. Significant developments are taking place in integrating data from sensor systems with information systems, and in improving the efficiency with which remotely sensed data interact with other data sets within such systems. NASA's goal in this work is not only to improve our ability to handle remotely sensed satellite data but also to improve our understanding of important biophysical and socioeconomic processes. To the extent that remote sensing expands the analytical potential and capabilities of resource managers and planners, it ought also to be a relevant concern in development of new information systems to support their decisions.

Finally, GIS technology is itself a developing technology, as is remote sensing — neither yet widely familiar or well understood within the research and user communities at large. This lack of understanding has in part served to isolate design of remote-sensing systems and techniques, hardware, and software from concepts of GIS design. As these technologies move into new stages of maturity — as many believe they are now doing — current problems of integrating the technologies will largely dissolve. As Shelton and Estes (1979) state, integration of remote-sensing and geographic information systems is not a matter of fundamental incompatibility of technologies, nor of the reluctance of technologists to collaborate. Integration of the technologies depends on realization that the potential of each cannot and will not be fully realized until they are efficiently and effectively integrated.

## *BIBLIOGRAPHY*


Anderson, James R., 1976. "Applications to Land-Use Mapping and Planning." *A New Window on Our Planet,* Geological Survey Professional Paper, U.S. Government Printing Office.

Anderson, J. R.; E. E. Hardy; J. T. Roach; and R. E. Witmer, 1976. "A Land Use and Land Cover Classification System for Use with Remote Sensor Data." U.S. Government Printing Office.

Calkins, H. W., and R. F. Tomlinson, 1977. "Geographic Information Systems, Methods, and Equipment for Land Use Planning 1977." U.S. Department of the Interior.

Colvocoresses, A. P., 1976, "Applications to Cartography." *A New Window on Our Planet,* Geological Survey Professional Paper, U.S. Government Printing Office.

Computer Sciences Corporation, 1978. LANDSAT and Ancillary Data Inputs to an Automated Geographic Information System: Applications for Urbanized Area Delineation, Contract No. NAS5-24350.

Computer Sciences Corporation, Knapp, E., 1978. "LANDSAT and Ancillary Data Inputs to an Automated Geographic Information System: Applications for Urbanized Area Delineation."

National Academy of Sciences, National Research Council, 1975. "Practical Applications of Space Systems — Part III, Landand Use Planning." NASNRC.

Shelton, R. L., and J. E. Estes, in press. "Integration of Remote-Sensing and Geographic Information Systems." *Proceeding of the Thirteenth International Symposium on Remote Sensing of the Environment,* Environmental Research Institute of Michigan, Ann Arbor, MI

Tom, C.; L. D. Miller; and J. W. Christenson, 1978. "Spatial Land-Use Inventory, Modeling, and Projection — Denver Metrotropolital Area, with Inputs from Existing Maps, Airphotos, and LANDSAT Imagery." NASA Technical Memorandum.

# Digital Image Technology Mapping, Charting and Geodesy Impact

*by Marshall Faintich*


## ABSTRACT

The Defense Mapping Agency Aerospace Center has developed a program to exploit digital image technology for the advancement of mapping, charting, and geodesy. Primary investigations include image processing, analysis and display techniques, and computer image generation. A dramatic impact has been made in the ability to analyze various digital data-bases produced by the Defense Mapping Agency by applying state-of-the-art digital image technology concepts. In addition, emerging technologies in image processing and digital image analysis for automated feature extraction are demonstrating the potential for both significant improvement in the utilization of imagery and reduction in production costs. Finally, computer image generation techniques are validating the content of both standard and proposed digital data-bases.


## INTRODUCTION

With the introduction of digital image technology into the scientific community, the Defense Mapping Agency Aerospace Center (DMAAC) began a proram to exploit this technology for the advancement of mapping, charting, and geodesy (MC&G). Primary areas of in-house application engineering are image processing, image analysis for automated feature extraction, image display techniques for data-base analysis, and computer image generation/sensor simulation for advanced weapon systems support. Specialized digital image technology hardware for production systems is continuing to be procured, but the major developmental hardware at DMAAC since 1974 has been the OPTRONICS Scanning, Digitizing, and Reimaging (SDR) System (Figure 1).

## IMAGE PROCESSING

To support both image analysis and electro-optical correlation systems, DMAAC has developed a general purpose digital image-processing software package to accomplish image restoration, enhancement, noise removal and geometrical warping. Spatial convolution filters, pixel histogram manipulations, transform coding and filtering, and compression techniques are included. Consider the following example using a LANDSAT image containing transmission noise (Figures 2 through 5). A 1 x 3 pixel median filter was applied to remove the noise. This was followed by a histogram equalization procedure to enhance the contrast, and then image warping was applied to rectify the image to the proper geometry.

A more powerful technique for enhancement is shown in Figures 6a and 6b. Here a 5 × 5 convolution matrix was developed to model a second order Laplacian enhancement with variable weighting factors. The original image is of the Martian surface taken from the VIKING orbiter spacecraft. The enhanced image display optimizes both gray level improvement and edge sharpening. A variation of this technique can be used for edge-only displays, as shown in Figures 7a and 7b. A number of Laplacian-enhanced and edge-only images have been used to support various types of electro-optical correlation systems.

## IMAGE ANLAYSIS FOR AUTOMATED FEATURE EXTRACTIONS

Some of the major programs at the DMAAC are the production of digital data-bases that describe the physical appearance of the surface of the earth. These data-bases include, but are not limited to, terrain elevation, culture including landscape characteristics, and vertical features. A major problem in the production of these digital data-bases is the extreme cost of primarily manual photo interpretation of surface features. Various digital techniques are being investigated to develop a digital image analysis capability for feature extraction. The techniques are addressing automated classification from both multi-spectral and conventional black and white imagery.

*Figure 1. Optronics SDR Equipment*

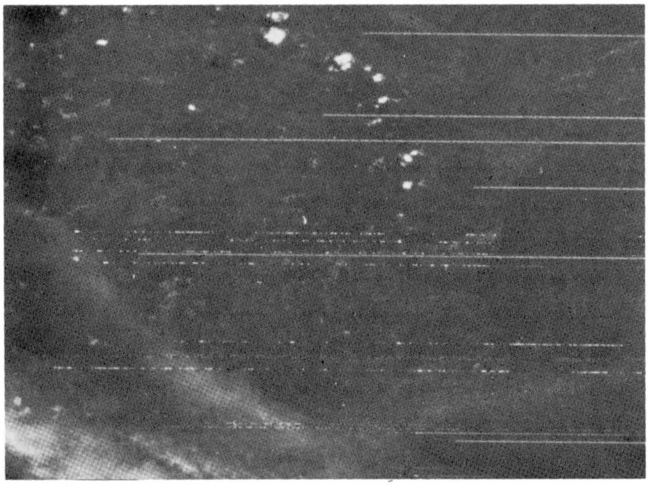

*Figure 2. Original LANDSAT Image with Noise*

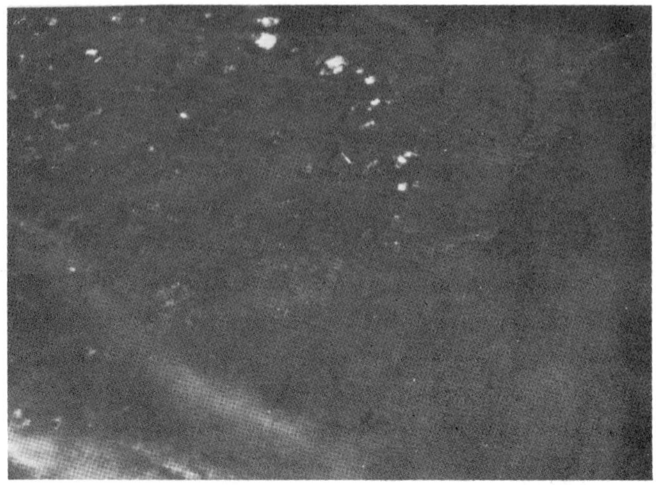

*Figure 3. Median Filtered LANDSAT Image*

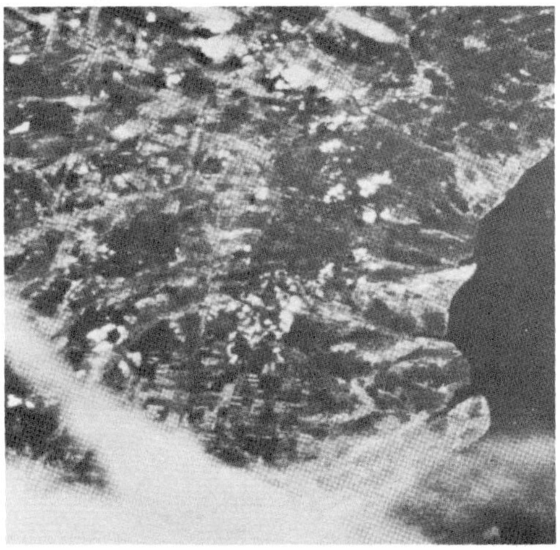

*Figure 4. Histogram Equalized LANDSAT Image*

*Figure 5. Rectified LANDSAT Image*

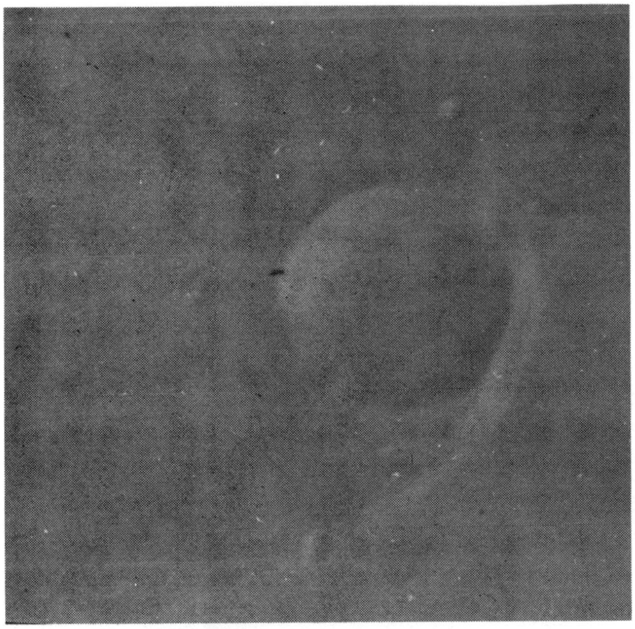

Figure 6a. Original VIKING Mars Image

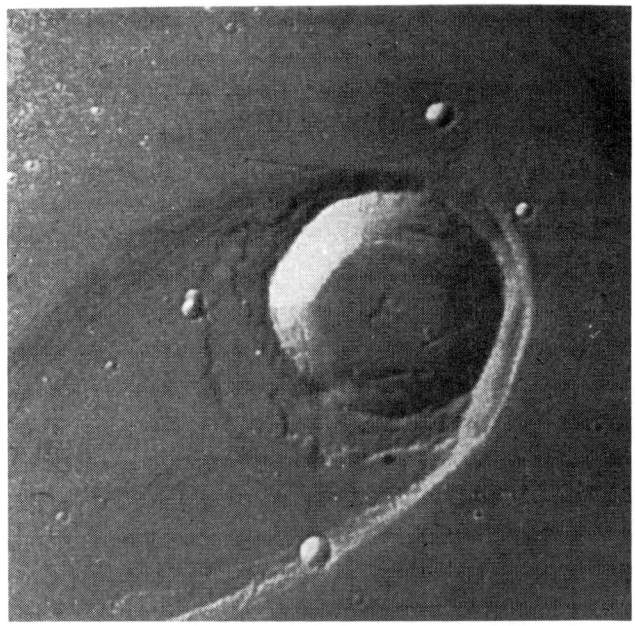

Figure 6b. Second order Laplacian-Enhanced VIKING Image

Figure 7a. Original Image

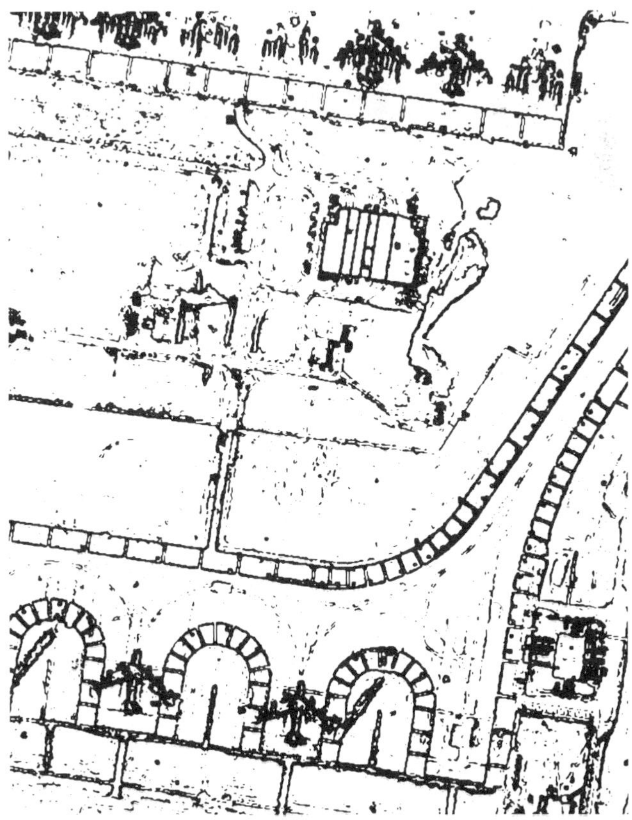

Figure 7b. Edge Display

Classical multi-spectral analysis using LANDSAT imagery has been investigated. Figure 8 shows the two principal components of the four multispectral LANDSAT bands over Phoenix, Arizona, after both principal component analysis and histogram equalization were applied. Figures 9a and 9b show automated vegetation and rock/concrete classification from these two components.

A more difficult problem concerns automated classification from black and white imagery. Current R&D contracts are considering this problem. Preliminary results were obtained (Faintich, 1977) by considering statistical measures of the pixel gray levels within small windows about each feature. Figures 10a and 10b show 1:110,000 scale black and white imagery over Blois, France, and the corresponding classification of wooded areas. Note that other feature types "leaked" through the process. Newer techniques are promising a better than 90% classification accuracy.

*Figure 8. Principal Components of LANDSAT Image*

*Figure 9a. Vegetation*

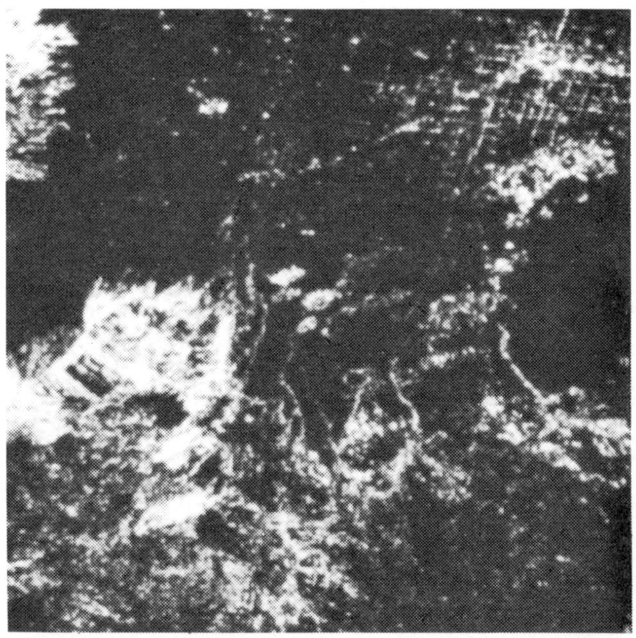

*Figure 9b. Rock/Concrete*

*Figure 10a. Blois Original*

*Figure 10b. Wooded Areas*

## *IMAGE DISPLAYS FOR DATA BASE ANALYSIS*

The digital terrain elevation data produced by the Defense Mapping Agency support a wide variety of products, including input to electro-optical sensor simulators, guidance systems, and automated cartographic systems. These data are collected from source maps using the Lineal Input System (Sippel, 1975), and from stereo-pairs of photographic imagery using optical correlation processes. In order to facilitate the display of these data-bases for analysis, techniques for various types of display have been developed (Faintich et.al., 1978a, 1978b).

Along classical MC&G lines, a digital terrain elevation matrix may be used to generate a standard contour plot (Figure 11) and the corresponding tint plate (Figure 12). Experimental cartographic products are using these techniques. For advanced analysis, terrain data may be depicted in the form of gray level elevation code or shaded relief from various sun angles (Figure 13). Additional information is gained from applying digital image convolution techniques to the digital terrain elevation data. Figures 14 and 15 show a shaded relief representation and a corresponding convolution of the data. Note the delineation of cartographically versus photogrammetrically derived data. Convolution techniques allow for one meter level analysis of the digital elevation data. In addition, photogrammetric models may be applied with shaded relief techniques to generate pseudo-stereo-pairs from the data-bases (Figure 16).

## *COMPUTER IMAGE GENERATION/SENSOR SIMULATION*

The primary objective of the digital sensor simulation investigations being conducted at DMAAC is to establish an editing and analysis capability for the digital culture and terrain data-bases. These data-bases are being produced by DMAAC to support advanced aircraft simulators by providing an improved low level radar training capability offered by the digitally generated radar landmass images. As a result of the technology developed for the aircraft simulator support, sensor guidance reference scenes are also being generated. In addition to radar scenes, visual and microwave scenes are being digitally generated. Earlier papers (Faintich, 1974, 1976a, 1976b) have detailed initial capabilities in scene generation. Currently, intensive studies are under way to generate synthetic input data bases with apparent resolutions finer than in the original data-bases, using supportive data-base information. Highly realistic sensor simulations have been generated, and the continuing emphasis is on both modeling new sensors and improving resolution without increasing data-base production costs.

The scenario for digital radar simulation is to first transform the off-line digital terrain and culture data-bases into an on-line terrain and radar reflectance potential data-base over a desired gaming area. The on-line data-base may be displayed for analysis as a computer generated synthetic image prior to sensor simulations. With radar control settings, weather

*Figure 11. Contours Generated from Digital Terrain Data: 100ft. Contour Interval, 500ft. Index Interval*

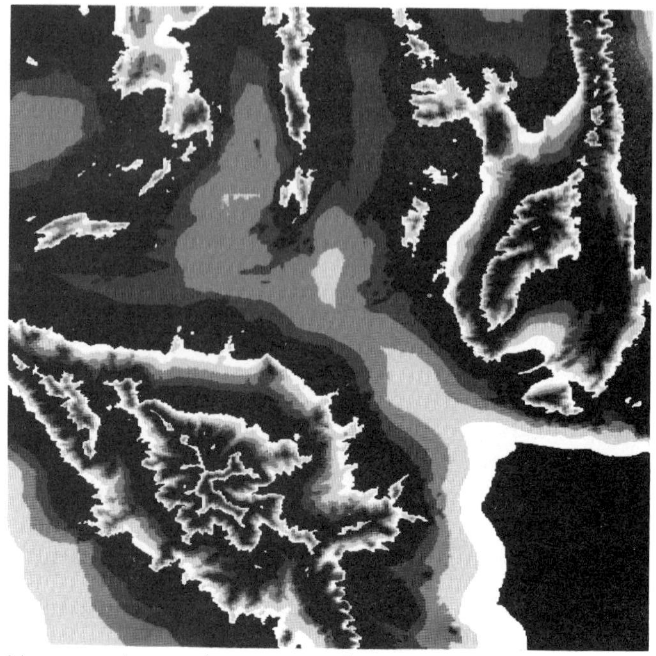

*Figure 12. 100ft. Contour Interval Tint Plate with Elevations Coded to Eight Shades of Gray*

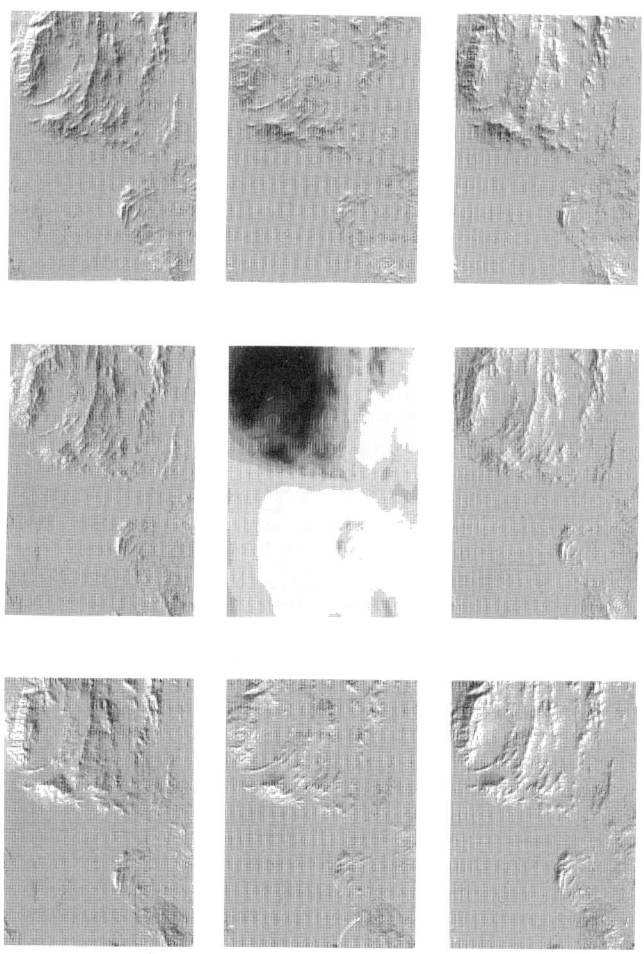

***Figure 13. Variable Sun-Angle Shaded Relief***

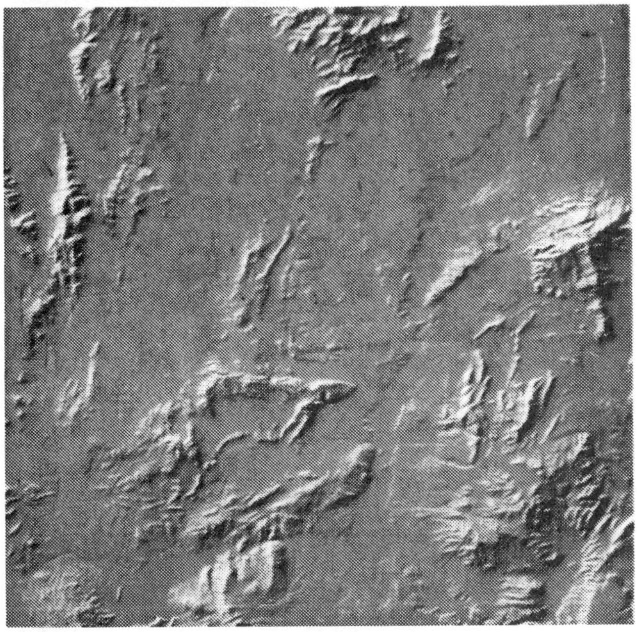

*Figure 14. Shaded Relief of Input Terrain Data for Convolution Studies*

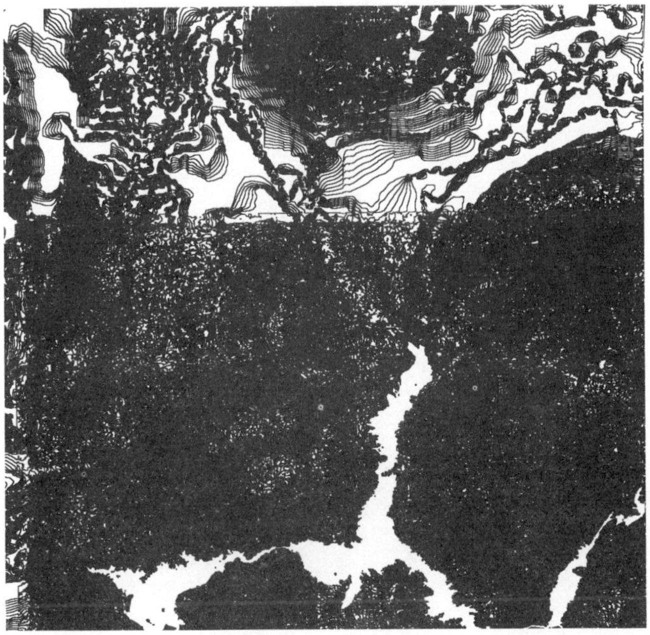

*Figure 15. Convolution Mask Results Delineating Cartographically versus Photogrammetrically Derived Data*

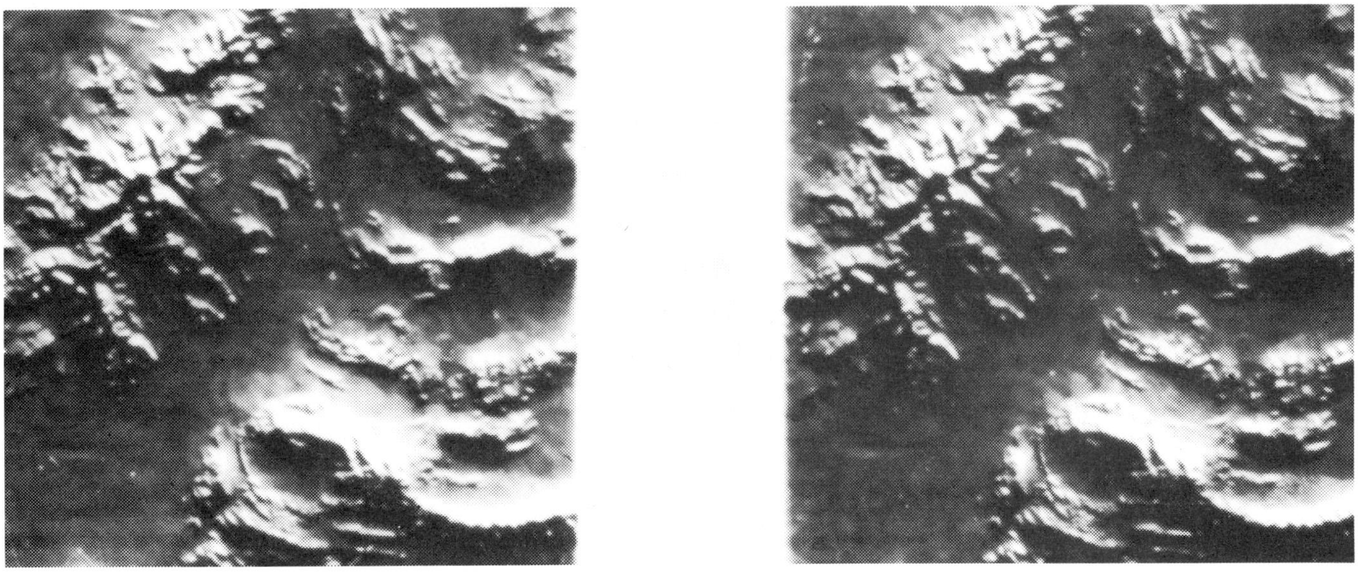

*Figure 16. Computer Generated Stereo-Pair*

parameters, and radar location as input, the software first generates an optical perspective geometry view of the observable ground area using predicted radar returns for the associated gray levels (Figure 17). This perspective view is then transformed to the desired radar geometry, and approximate radar effects are added to the simulation. The final output matrix is then buffered to a hard copy display device. Similar scenarios exist for other types of sensors, with minor changes in the on-line transformation, and sensor module changes to the simulation software.

The current standard production data-bases (Level I) correspond to cultural information displayed on a 1:250,000 scale chart and digital terrain data on a 3″ grid. Detailed information is available in "Product Specifications for Digital Landmass System (DLMS) Data-Base." In general, this level of information has proved to be adequate for long range (Figure 18) and medium range (Figure 19) radar simulation.

With the development of advanced high resolution radar and visual simulators, higher resolution cultural data bases are required. Production resources limit this higher resolution data

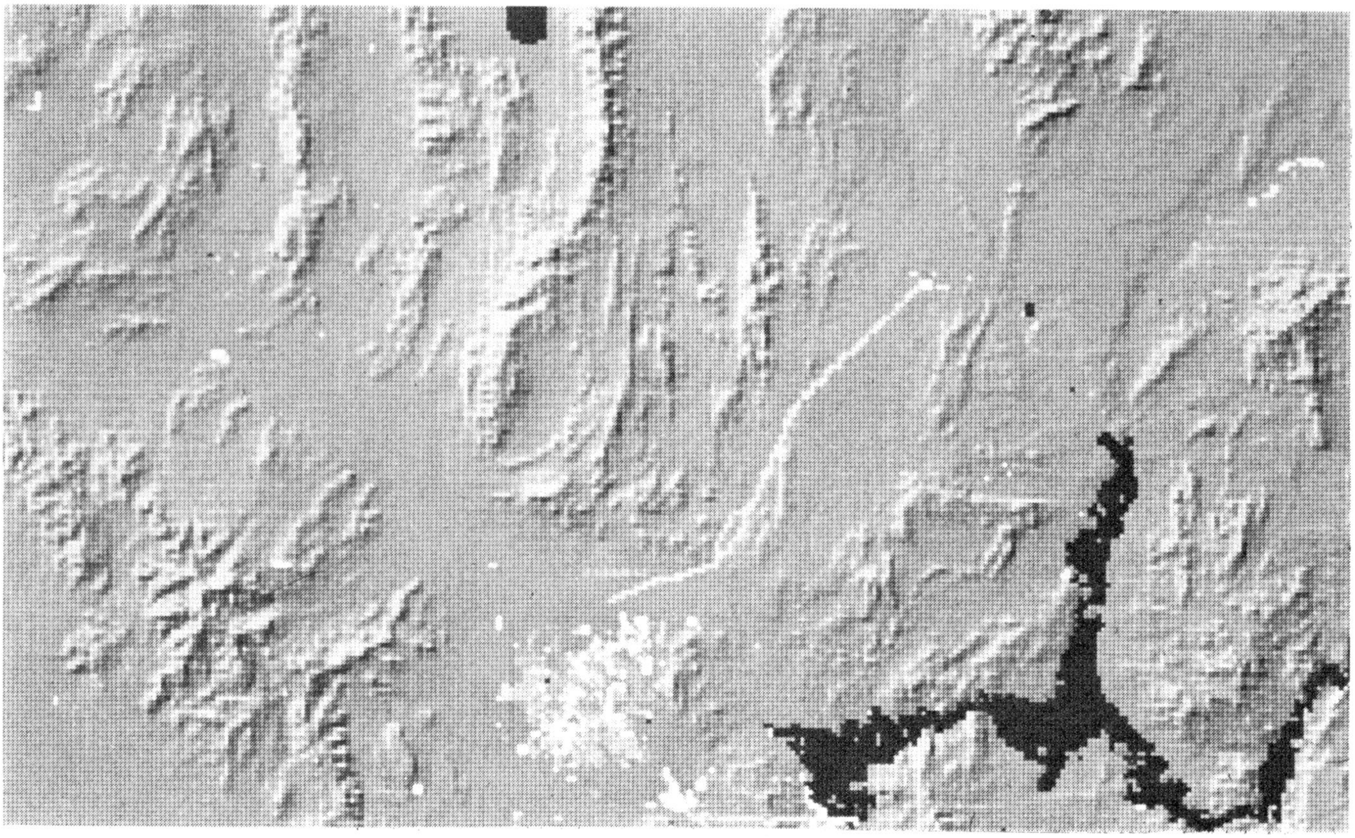

*Figure 17. On-Line Data-Base Display: Las Vegas, Nevada*

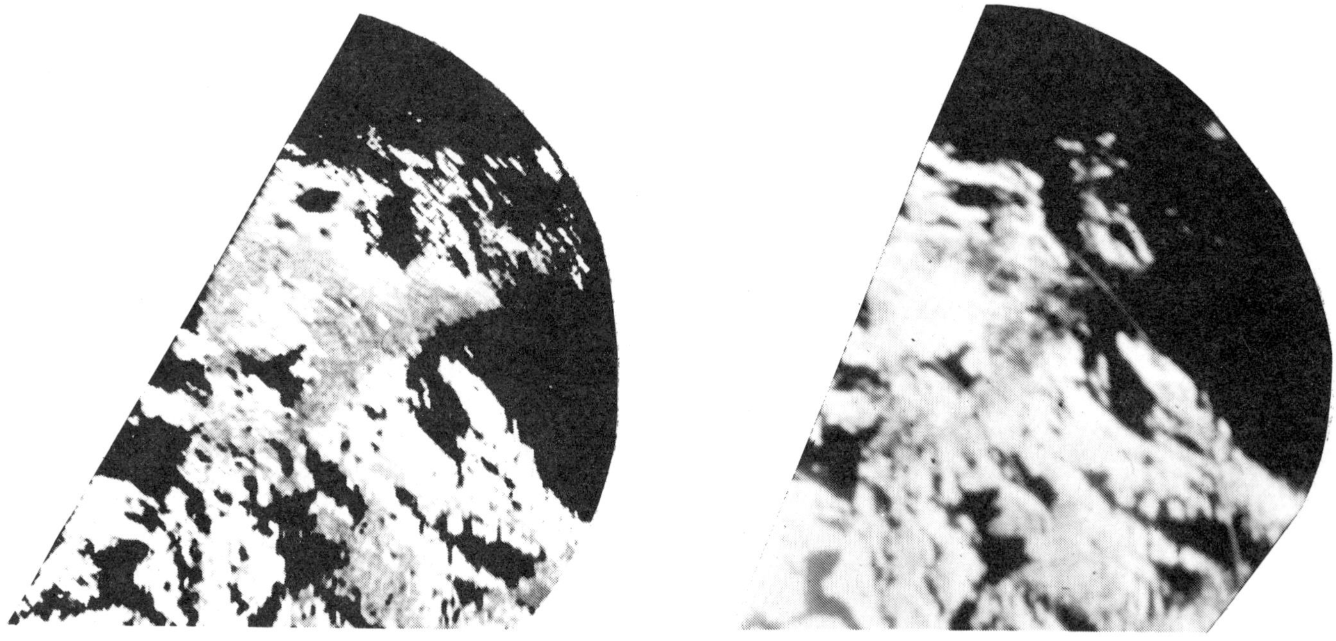

*Figure 18. Long Range (85NM) Radar Simulation (Left); Actual (Right)*

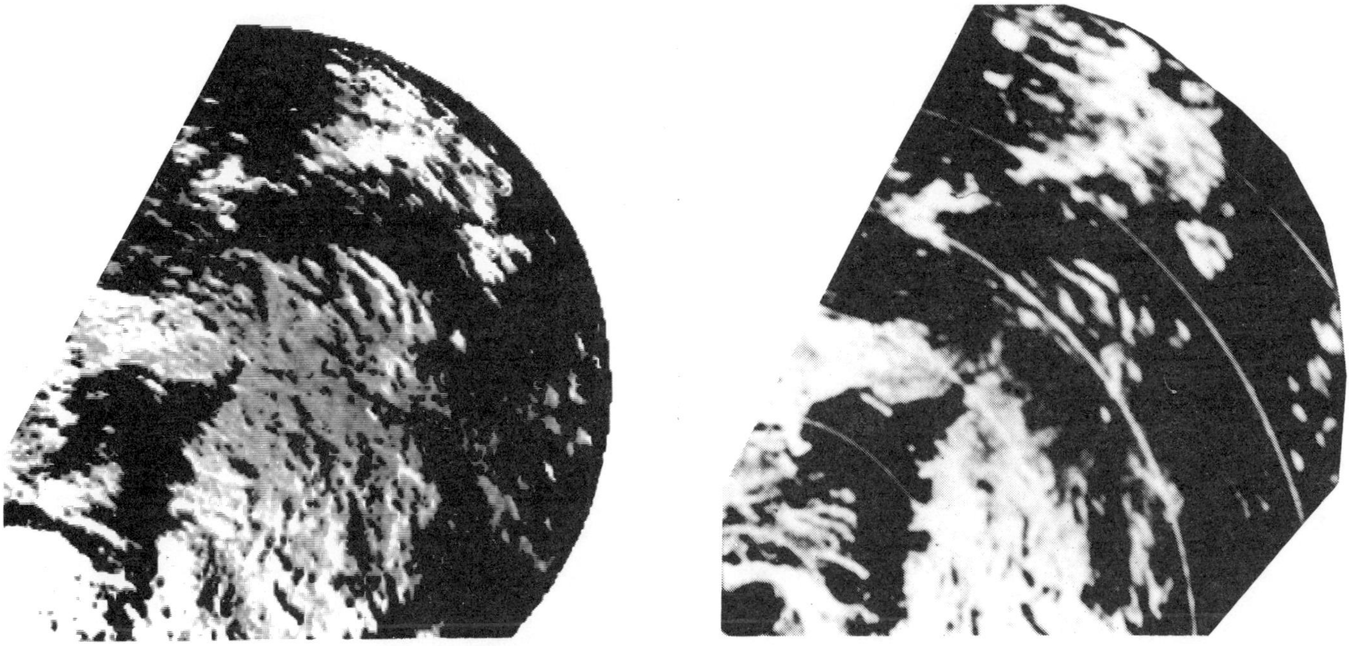

*Figure 19. Medium Range (32 NM) Radar Simulation (Left); Actual (Right)*

to small geographic areas of interest, and worldwide coverage is unattainable without automated feature extraction techniques. Due to the extremely high production cost for very high resolution off-line data-bases, efforts have been under way to simulate high resolution on-line data-bases from lower resolution off-line input, and then to generate the simulated sensor scenes from the synthetic data-base. Current results show a high degree of potential success associated with enormous cost savings and the appearance of greater resolution in the simulations.

The current high resolution data-bases (Level II) correspond to cultural information displayed on a 1:50,000 scale chart. The Level I data-base aggregates Level II type information into homogeneous areal features with associated predominant descriptors, including percentage tree and roof cover. The synthetic feature break-up technique generates random cultural, tree cover, and background features within the areal feature boundaries, based on the percentages of tree and roof cover. Cultural descriptors are then randomly assigned with a normal distribution about the predominant values of the original areal features. The Level II data also contain aggregate areal features of very high resolution, and the synthetic break-up process may be applied to any level of data.

This type of synthetic feature break-up allows for greater realism for Level I simulation as well as the blending of actual Level II data patches within a Level I data-base. Examples of both radar and visual simulations depicting the synthetic break-up technique are shown in Figures 20 through 24.

Additional investigations are being undertaken to refine gray level assignments, to add texture to both landscape and man-made cultural features, and to add new types of features to the data-base. The development of this sensor simulation capability has played a key role in the refinement of digital data-base production specifications.

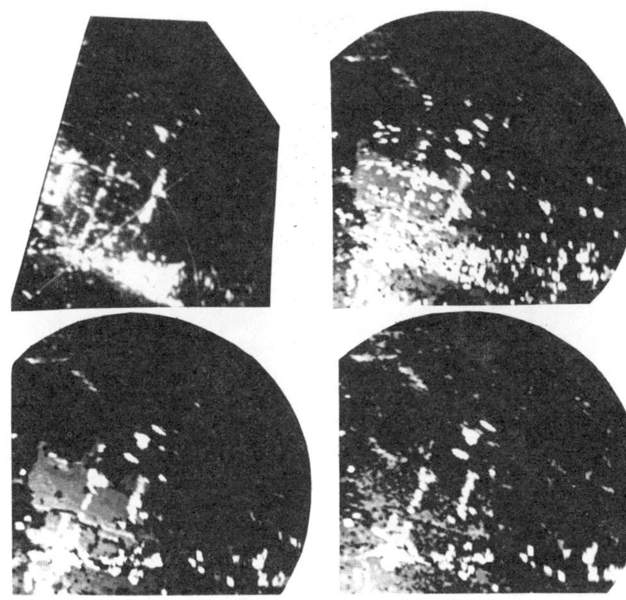

*Figure 20. Short Range (15NM) Radar: Actual (Top Left); Level II Simulation (Top Right); Level I Simulation (Bottom Left); Level I Synthetic Break-Up (Bottom Right)*

## CONCLUSIONS

The development of digital image technology has made a dramatic impact upon classical production methodologies for mapping, charting, and geodesy. DMAAC has begun to develop a new line of digital products to support advanced aerospace weapons systems, and digital image technology is playing a key role in the analysis of these products. In addition,

*(Cont'd. on page 40)*

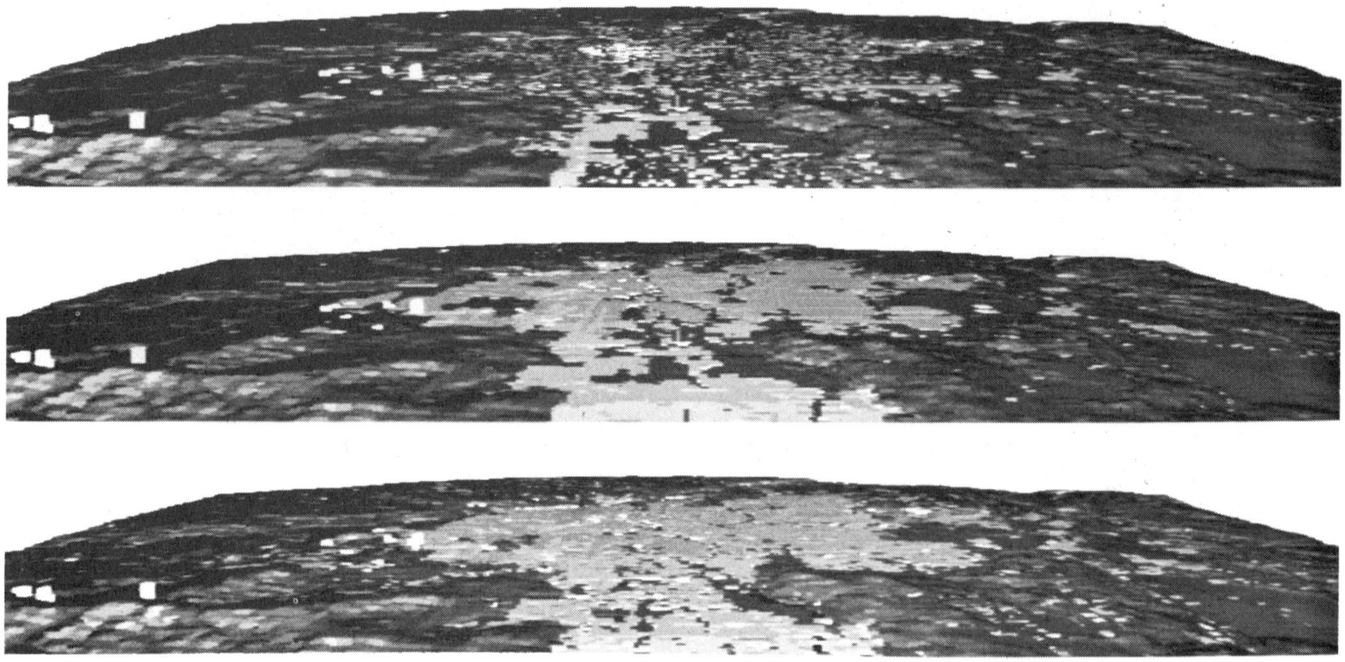

*Figure 21. Visual Simulations over Spokane, Washington (13,000 ft. AGL); Level II (Bottom); Level I (Middle); Level I with Synthetic Break-Up (Top)*

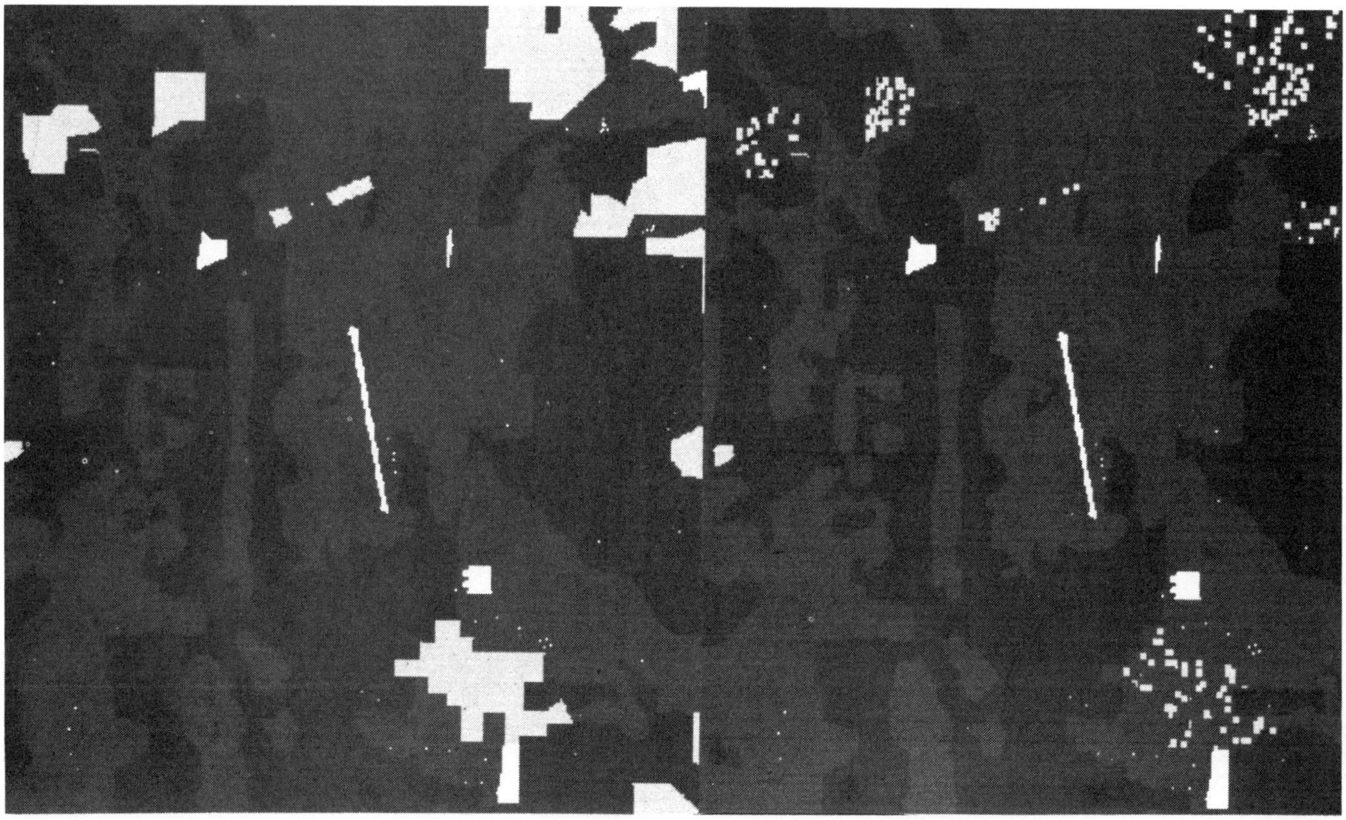

*Figure 22. Level I On-Line Visual Data-Base Display (Left); with Synthetic Break-Up (Right)*

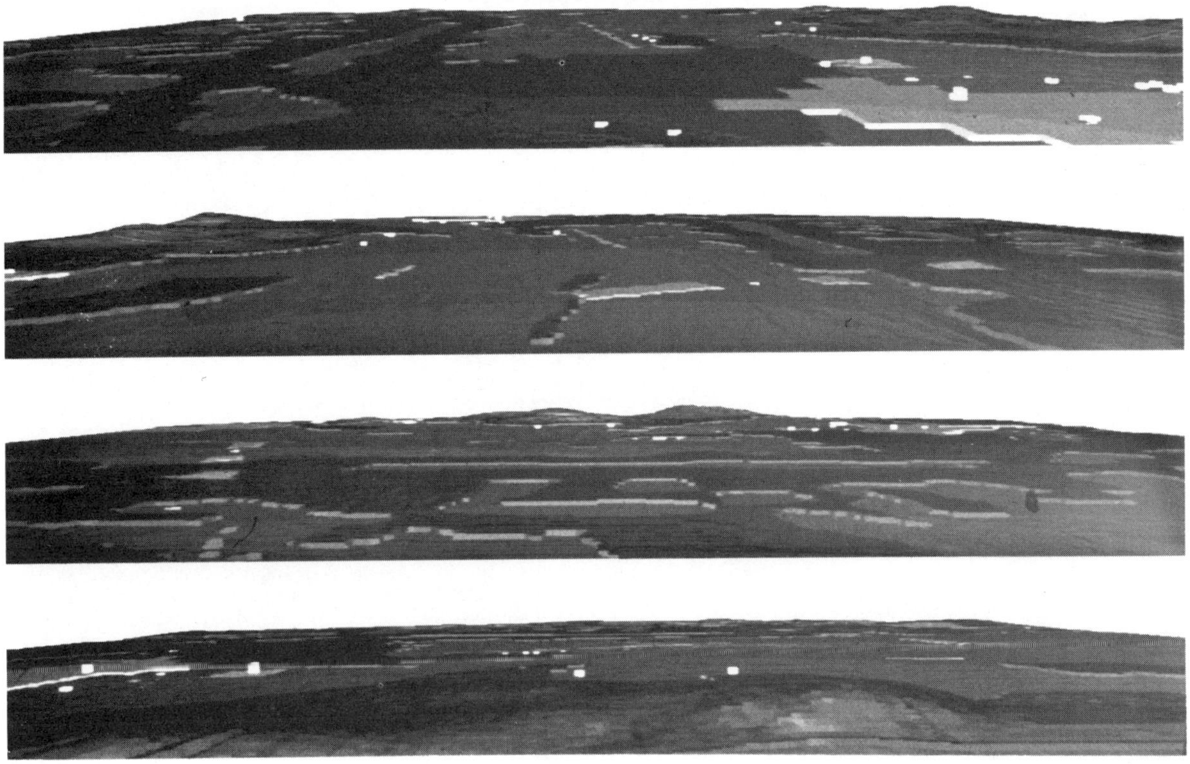

*Figure 23. Level I Visual Simulations (Top to Bottom: North, South, East, West); All Scenes Are 1500ft. AGL, 2.5 NM from Airfield*

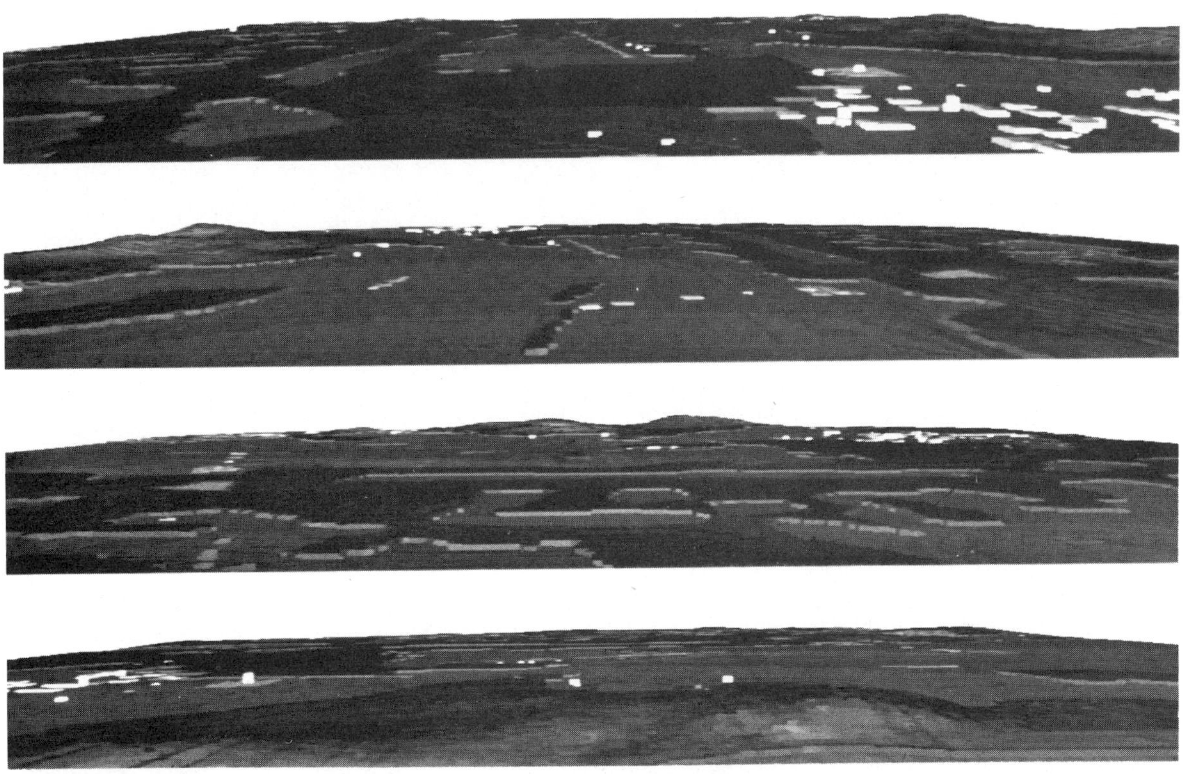

*Figure 24. Level I Visual Simulations with Synthetic Break-Up (Top to Bottom: North, South, East, West); All Scenes Are 1500ft. AGL, 2.5 NM from Airfield*

emerging technologies in image processing and digital image analysis for automated feature extraction are demonstrating the potential for both significant improvement in the utilization of imagery and reduction in production costs. Finally, computer image generation techniques are validating the content of both standard and proposed digital data-bases.

## ACKNOWLEDGEMENTS

The following individuals have made significant contributions to the material presented in this paper: Bernard Kolo, James Simpson, David Fahy, and George Sigler.

## REFERENCES

[1]Faintich, M. B., 1974. "Digital Scene and Image Generation." *Naval Weapons Laboratory Technical Report,* TR-3147.

[2]Faintich, M. B., 1976a. "Digital Sensor Simulation." *Photogrammetric Engineering and Remote Sensing,* 42, 11, pp. 1427-1440.

[3]Faintich, M. B., 1976b. "Advances in Digital Sensor Simulation." *Proceedings of the American Society of Photogrammetry,* Fall Convention, Seattle, WA, pp. 283-302.

[4]Defense Mapping Agency, 1977. "Product Specifications for Digital Landmass System (DLMS) Data Base." St. Louis Air Force Station, MO.

[5]Faintich, M. B., 1977. "Feature Identification Using Statistical Signatures of the Regional Meso-Structure." Institut Geographique National Technical Report, Paris, France.

[6]Sippel, J., 1975. "Automated Editing of Digital Graphic Data." *Photogrammetric Engineering and Remote Sensing,* 41, 8, pp. 1047-1058.

[7]Faintich, M., G. Sigler; D. Fahy, 1978a. "Digital Image Display and Simulation from Digital Terrain Data Bases." *Digital Terrain Model Symposium Supplement,* St. Louis, MO, 9-11 May 1978, pp. 610-624.

[8]Faintich, M.; G. Sigler; D. Fahy, 1978b. "Advances in Digital Image Display and Simulation from Digital Terrain Data Bases." *Proceedings of the International Society of Photogrammetry Symposium,* Commission IV, Ottawa, Canada, 2-6 Oct. 1978.

[9]Faintich, M.B., 1979. "Digital Sensor Simulation at the Defense Mapping Agency Aerospace Center." *Proceedings of the National Aerospace and Electronics Conference* (NAECON), Dayton, Ohio, 15-17 May 1979.

# A Case Study in Computer Mapping in Petroleum Exploration A Geological Study of the Powder River Basin

*by David R. Feineman*

## *PROBLEM DESCRIPTION*

The Powder River Basin study was done to see if computer processing could help evaluate the petroleum exploration possibilities of a geographic area. As domestic petroleum supplies decline, new techniques must be developed to find oil in frontier areas, as well as to generate new plays in mature producing areas. This study focused on computer mapping of a developed area. All data were commercially available.

A geologist normally generates oil and gas prospects by evaluating information from surrounding wells. Considerable time can be spent in file searches, data selection, tabulation, and plotting of raw data before starting any interpretation. From this information the geologist will produce graphic displays, such as cross-sections, fence diagrams, and contour maps, to aid in his analysis of the data. In fact, 90 percent of exploration data are frequently presented in graphic form. If we are to effectively meet the explorationist's needs in any geological study, we must be able not only to replicate output that he would usually do by hand, but also be able to manipulate the data and generate new types of output that it would be impossible or impractical to do by hand. Computer mapping is a tool that allows us to handle large volumes of data and then merge it into a model of geologic history.

Underlying these requirements of geological computer applications are two critical factors. The first is that the geologist's interpretation of an area must be incorporated into the mapping process. Regardless of how well designed computer gridding or contouring algorithms are, there will still be cases where an unnatural looking or improperly extrapolated area will be produced. Computer mapping is only accepted if such spurious features are eliminated by grid and/or data modification, and changed to reflect the geologist's interpretation. The second critical requirement of computer mapping is that output must be available in a meaningful time frame.

The Powder River Basin computer mapping project had three basic goals. First, exploration management had expressed an interest in reexamining the exploration possibilities within the Paleozoic section of rocks in the Powder River Basin. A suite of detailed maps would aid in this analysis, and the basic contour maps could be augmented by additional graphics, such as isometric diagrams to help demonstrate the structural relationships between horizons. Second, we wanted to evaluate just how practical computer file retrieval and mapping from commercial files could be for the analysis of basins. Third, the project would benchmark Phillips Petroleum's processing techniques/software, and data-bases. Specific issues were an evaluation of: 1) the value of commercial well data files for regional studies; 2) Phillips' in-house retrieval software; 3) Phillips' mapping software and its ability to handle the large volumes of control point and cartographic data available for such studies; 4) the manpower required to do a regional basin evaluation; 5) the cost of such methods; and 6) a generalized procedure that could be set up to handle future studies.

The request for the Powder River Basin study was initiated by the exploration management of the Natural Resources Group of the Phillips Petroleum Company. The work was performed by the Computer Applications Unit of the Exploration Projects Section. This unit of geologists and technicians is responsible for applications processing of exploration data, training, and long-term applications development and coordination.

## *METHOD DESCRIPTION*

The problem was approached in three phases. Phase one was a search of digital bibliographic files to retrieve articles related to the geology of the basin. The second phase was the generation of an inventory of the digital well data, and a summary of reservoir and field data. Phase three was the actual computer mapping of well data.

For each horizon mapped, data were retrieved from the master well data file and loaded on an on-line disk file. DIMECO state and county lines, CIA World Data Bank-II rivers, and a Phillips propietary township and range base were also placed in on-line subfiles. Data sets were merged in the mapping phase and composited on the plotter. Extensive use of disk data sets allowed faster access times and facilitated the development of foreground processing programs to detect and eliminate errors in the the raw data.

Because a significant portion of the execution time of any contour mapping system is expended in gridding, the grid for each horizon was dumped to disk as a binary unformated file. Successive mapping runs to produce isometric diagrams, profiles, etc. loaded the saved grid file and then began graphics processing. An unacceptable surface was reprocessed using masks to control which subareas of the grid should be revised. In this manner, a set of data points was gridded only once and was then patched and filtered until it produced a surface that met all interpretive criteria.

We knew that the commercial file had inaccuracies in well locations and the elevations of horizons. Although work maps were scanned for obvious discontinuities in the data, no other precautions were taken to validate the data. It was hoped that on a regional scale, minor irregularities in the data would not be evident. As will be seen in the graphics examples, obvious data errors were not nearly as important as the lack of consistent relationships between surfaces. Since there is no consensus between geologists on "picking" the top of a particular rock formation, we suspected that errors in depth would be volumetrically more critical than errors in location. An attempt to compensate for the "noisy" data was made by using two distinctly different weighting functions in the piecewise linear least squares gridding algorithm. The first was a strictly deterministic weighting curve, that assigned an infinite weight to a control point and tended towards a linear trend fit away from control points. The second was a "statistical" weighting function that did not yield extremely large weights at control points. By using this smooth curve, more control points make a contribution to surface fit rather than having a few control points dominate. A grid produced in this way is the product of extensive averaging of the input data. This type of weighting unfortunately tended to obscure too much of the overall basin structure and was not actively pursued.

## *RESOURCES REQUIRED*

None of the hardware used to produce the graphics in the Powder River Basin study was specifically tailored to this application. Phillips has three IBM 370/168 mainframes with an attached processor. Batch mapping runs and data editing functions were performed on IBM 3277 or 3278 terminals operating under TSO. A Tektronix 4014-1 terminal was available to display previews of graphics and for interactive mapping runs. Plot tapes were received at a Harris RJE station, which consisted of a line printer, card reader, and tape drive. A Calcompt 748 flatbed plotter (which has since been replaced by a Xynetics) and a 42 inch Versatek electrostatic plotter were used off-line to produce hard copy maps.

### Software

The retrieval software to access the well file and World Data Bank 2 were developed in house, as was the program to generate cross-sections. All other graphic displays for this presentation were generated by CPS 1 (Contour Plotting System), a software package developed by Unitech, Inc. in Austin, Texas. CPS has all the standard mapping system capabilities, plus the ability to handle faults, generate confidence surfaces, make gradient vector displays and isometric contour plots.

### Data-Bases

The cartographic data used in this project came from several sources. State and county lines were obtained from the DIMECO file. Rivers were extracted from the CIA's World Data Bank 2. Township and range boundaries were digitized in house from USGS topo sheets. Proprietary software was available to plot the stored townships, and to calculate the location of a well based on its legal description.

The only source of well data was Petroleum Information, Incorporated's Well History Control System. This commercially available file contains detailed data on more than one million domestic oil and gas wells and was developed in cooperation with the petroleum industry.

## *PROBLEMS*

As was noted before, obvious errors in the control point data set were found by producing a series of check maps. A series of simple interactive programs were developed to access the well subfiles and retrieve all wells within a specified latitude and longitude and depth range. In this way, a technician could easily locate the bad control points and delete them from the data set.

The study turned up a number of problems that remain unresolved. How can you evaluate the errors in a commercially obtained file? What level of contour generalization is best for a regional map? Part of the answer may be available to us from the mapping system. During gridding, we can generate a confidence grid — a variance surface that allows us to determine the relative accuracy of any part of the grid. By attempting to analyze the variance surface, we might flag data busts as well as areas where the actual structure was more complex than the surface we modeled in gridding.

In the final analysis, much of the problem in contour mapping centers around grid generation and modification. Speeding up the iterative process of updating grids is a crucial point in any contour mapping problem. When the raw data to be mapped is over sampled and noisy, the first pass through the data will probably not be an adequate model of the surface. Not until the gridded surface array conforms to our interpretation can we produce meaningful graphics.

## GRAPHICS EXAMPLES

| Type | |
|---|---|
| Cross-Section | *Description:* Two-dimensional plots showing the change, from control point, of some mappable boundary by straight lines. They can be datumed on various horizons to aid in the interpretation of geologic history by approximating structure at the time of deposition. Generally produced early in a study for consistent correlation of horizons to be mapped.<br>*Advantages:* Simple to construct. Do not require any grids. |
| BASE MAP | *Description:* A map showing cartographic base data without any computer contouring. Files displayed are: DIMECO state/county lines, World Data Bank-II rivers, lat/log tick marks, and township and range lines. Control points could also be posted on the base.<br>*Advantages:* Defines the area to be mapped. Data requiring considerable interpretation to be mapped can be posted, freeing the geologist from the map drafting chore. |
| STRUCTURAL CONTOUR MAP | *Description:* A type of isopleth map. Structure contours connect points of equal altitude on a geologic surface. The map is made by interpolating between the elevation of the horizon at drill holes.<br>*Advantages:* Allows the interpretation of regional trends in the data. Surface structures, such as troughs and domes, and discontinuities can be identified. |
| ISOMETRIC DIAGRAM | *Description:* Three-dimensional block diagram of a structure surface. Viewing direction, angle, and vertical exaggeration are all under user control.<br>*Advantages:* Simplifies the interpretation of complex structures. Can easily be understood by nontechnical personnel. |
| RAISED CONTOUR DISPLAY | *Description:* An isometric display in which isopleth lines are raised in proportion to the value of the contour. Hidden lines are removed.<br>*Advantages:* Combines the ease of interpretation of an isometric diagram with the quantitative basis of an isopleth map. |
| RAISED CONTOUR AND CEILING CONTOURS | *Description:* A raised contour display overlaid by a standard contour map skewed into isometric space. Structural nuances evident on the base can readily be related to the original contour map. Cartographic base data can be plotted on the upper plane.<br>*Advantages:* An impressive display for management. Easy to interpret. Suitable for technical and nontechnical displays. |
| ISOMETRIC DIAGRAM AND CEILING CONTOURS | *Description:* A standard "fishnet" isometric overlaid by a skewed floating plane contour map.<br>*Advantages:* The three-dimensional impression is slightly enhanced by using an isometric base. |
| STACKED CONTOUR MAPS | *Description:* The structure contour maps from multiple horizons are skewed into isometric space and stacked above one another. An isometric or raised contour display could form the base of the stack. Cartographic base data can be displayed on any of the planes, or on a separate layer.<br>*Advantages:* Multiple surfaces can be displayed and interpreted simultaneously. Changes in structural pattern become readily identifiable, as do inconsistencies between layers. |
| GRADIENT VECTOR DISPLAYS | *Description:* The structure surface is differentiated into two components: magnitude and direction. Vectors whose lengths are proportional to the surface magnitude and directions are controlled by slope and then drawn from grid nodes.<br>*Advantages:* Useful for picking out discontinuities within a data set. |

## RESULTS

Using a well data file and the computer, a single geologist could rapidly map and summarize information on the exploration potential of a basin. Data for such a study could easily be prepared for management presentation within a month, in a concise and easy to interpret form. By comparison, an area of only 5,000 wells mapped manually could easily take six months and would lack any type of three-dimensional displays. Computer mapping had extended our capability to evaluate much more data, and to display it in a high impact form that management responded to.

Impressed by the data and the high rate of turnaround on this project, exploration management immediately requested regional study of the Big Horn Basin. The technical staff was not prepared for the prospect of beginning another high priority project immediately, and despite the experience gained in the Powder River Basin, this new study never received the interest or acclaim of its predecessor.

The project was not completed without considerable manpower requirements. It is estimated that the four people working expended 328 man/hours to complete the job. Future studies should require far less time, since many file manipulation techniques were being tested. The maps presented here were created specifically to show the variety of computer graphics available. All maps were created in less than one week by one geologist.

## CONCLUSIONS

The Powder River Basin project was a success. As you may recall, its specific objectives were to evaluate: 1) the value of commercial well data files for regional studies; 2) Phillips' in-house retrieval software; 3) Phillips' mapping software and its ability to handle the large volumes of control point and cartographic data available for such studies; 4) the manpower required to do a regional basin evaluation; 5) the cost of such methods; and 6) a generalized procedure that could be set up to handle future studies. This study demonstrated the following key points.

Commercial well data files are a readily available source of mappable information. Posted base maps can be made quickly, with little difference in quality over more costly manual searches of company files. A sophisticated data base management and retrieval system is not a prerequisite to performing such tasks. Large volumes of data produced by searching broad areas are not a problem — in fact they make computer mapping more desirable as manual drafting manpower estimates skyrocket. Some commercially available mapping packages, such as CPS 1, are theoretically capable of handling infinitely large data sets. Our experience has shown that we can economically afford to grid and contour a 5,000 point data set. Using computer graphics, we can go beyond standard contour mapping, and provide additional isometric displays that help both manager and explorationist visualize a prospect. Computer mapping is a cost effective alternative to manual drafting that can aid in our understanding of the complex spatial relationships between lithologic units. The techniques described here could be applied to any high control point density mapping problem. The use of isometric contour and ceiling plots seems to be a good management presentation tool that could be used as effectively outside the petroleum industry.

## REFERENCES

Dillon, E. L., 1967. "Expanding Role of the Computer in Geology." *American Association Petroleum Geologists Bulletin,* v. 52, p. 1185-1201.

Forgotson, J. M., Jr., and P. H. Stark, 1972. "Well-Data Files and the Computer, a Case History from the Northern Rocky Mountains." *American Association Petroleum Geologists Bulletin,* v. 56, p. 1114-1127.

Forgotson, J. M., Jr., and J. L. Stout, 1969, "Future of Well-Data Information Systems." *Computer Applications in the Earth Sciences — An International Symposium,* Plenum Press, NY, p. 61-72.

Harbaugh, J. W., and D. F. Merriam, 1968. *Computer Applications in Stratigraphic Analysis,* Wiley and Sons, p. 10-29.

Kaplan, A. E.; R. W. Gimlan; R. J. McGlew; and F. P. Pratt, 1978, "Computer Applications in Regional Study of Powder River Basin." *American Association Petroleum Geologists Bulletin,* v. 62, p. 528.

Rockwell, D. W., and R. Roberts, 1973. "An Application-Oriented Exploration Data-Base System." *Oil and Gas Journal,* v. 71, no. 8, p. 62-78.

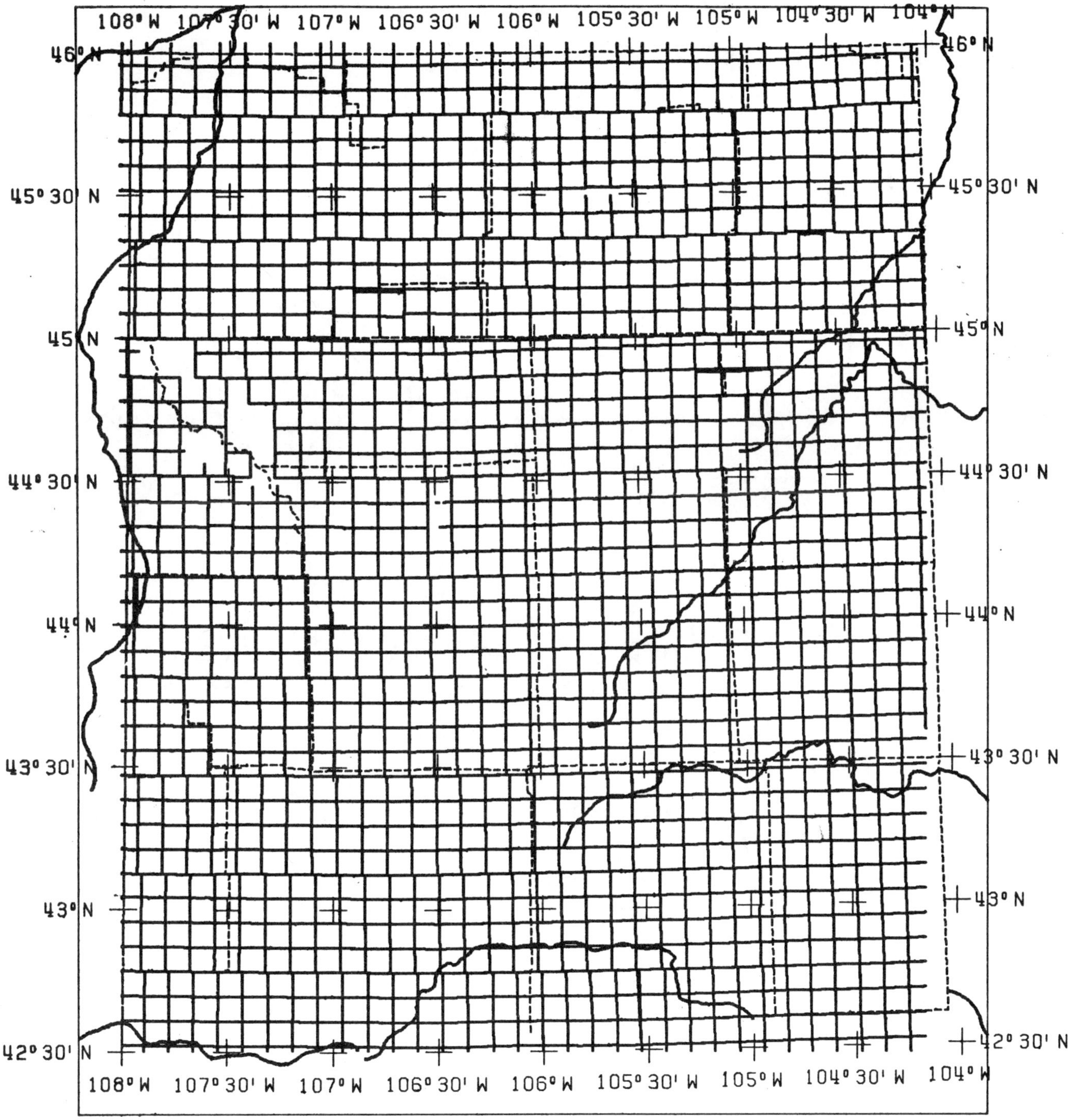


*Figure 1. Base Map*

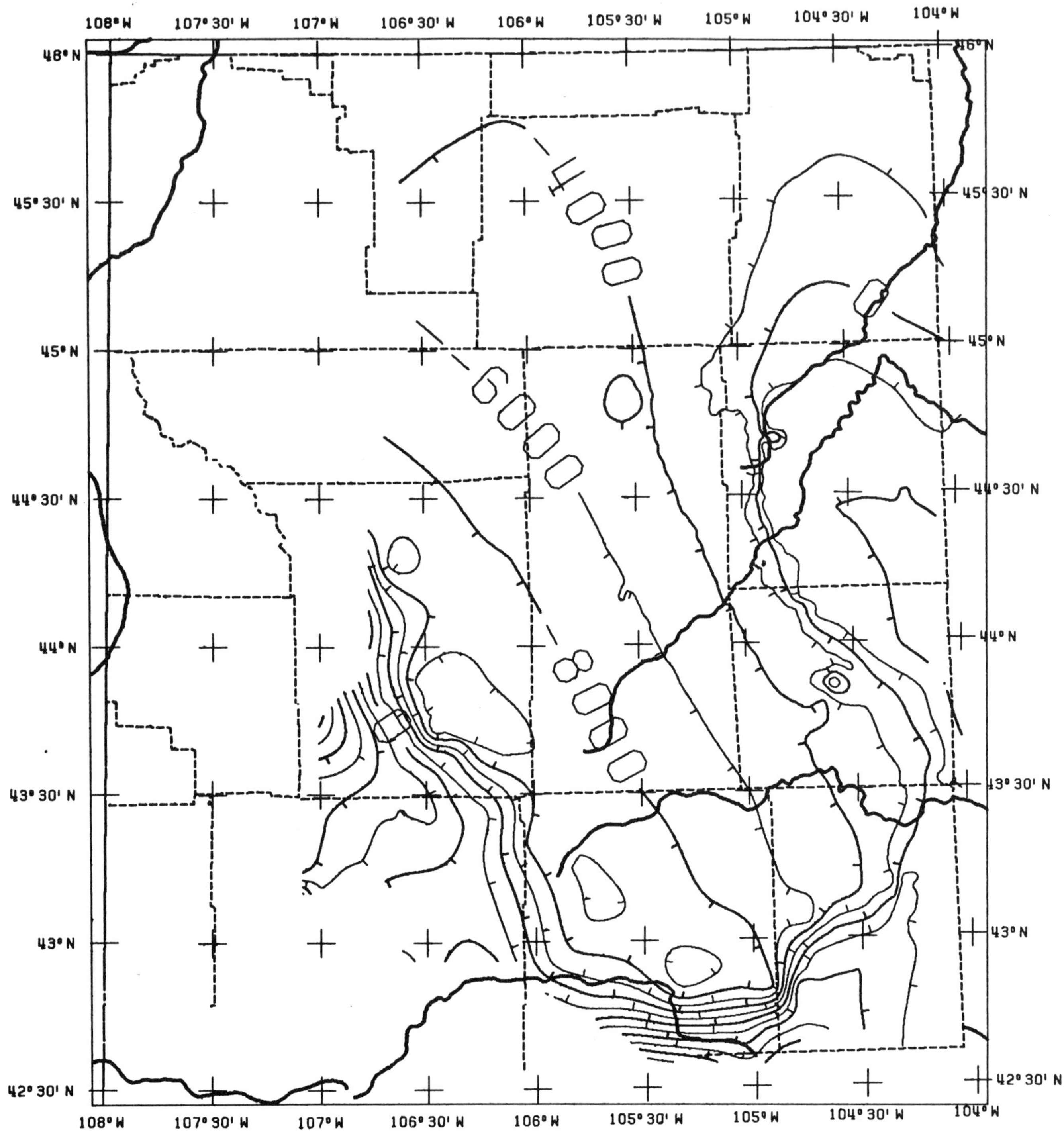


*Figure 2. Structure Contour Map (Minnelusa Formation)*

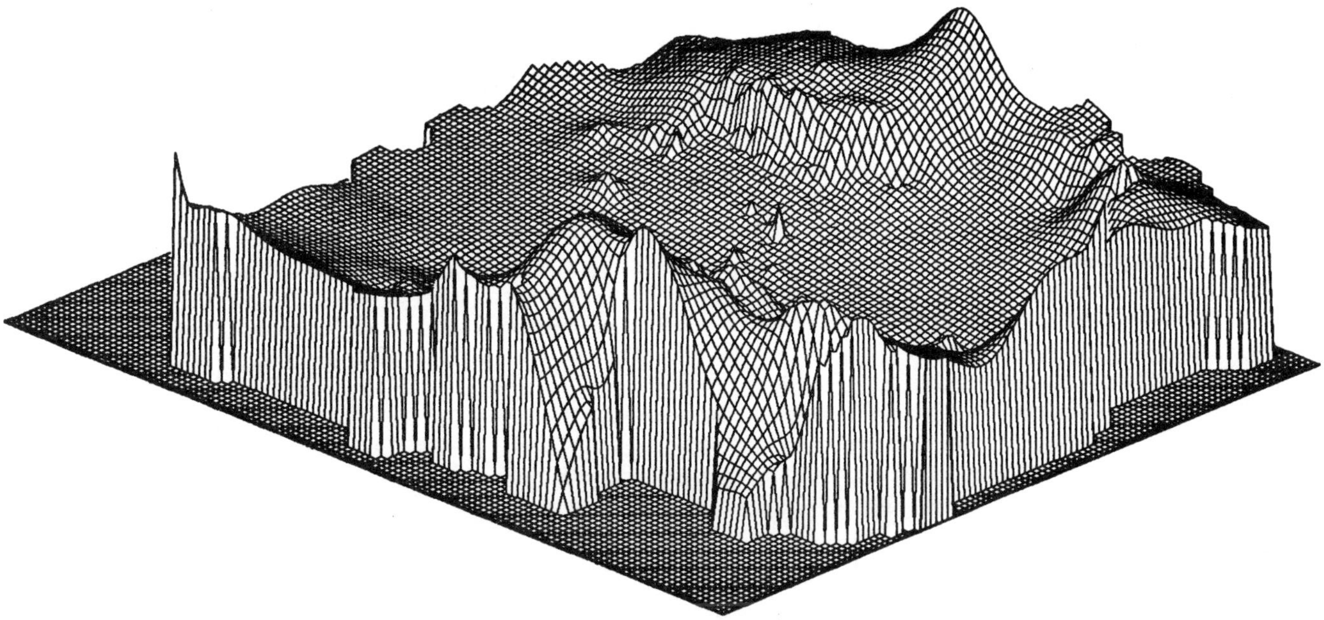

*Figure 3. Isometric Diagram (Mississippian Structuro)*

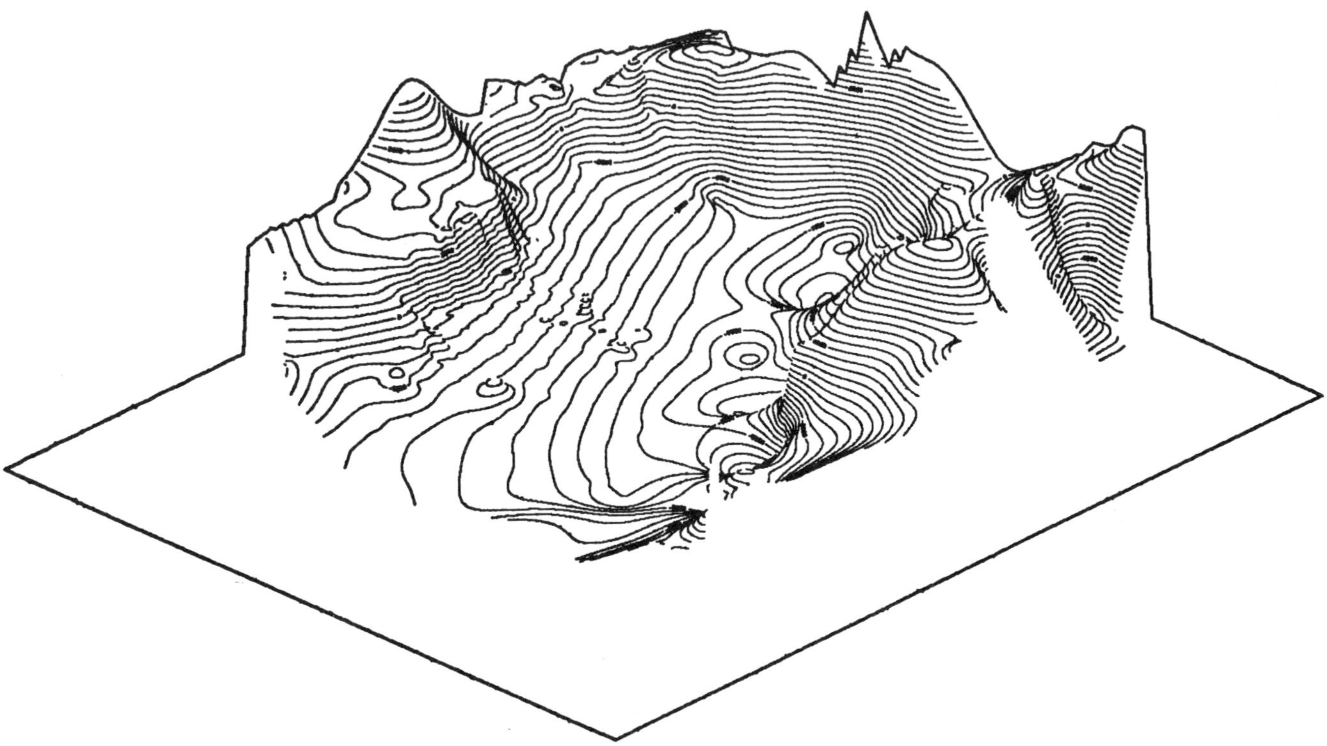

*Figure 4. Raised Contour Display*

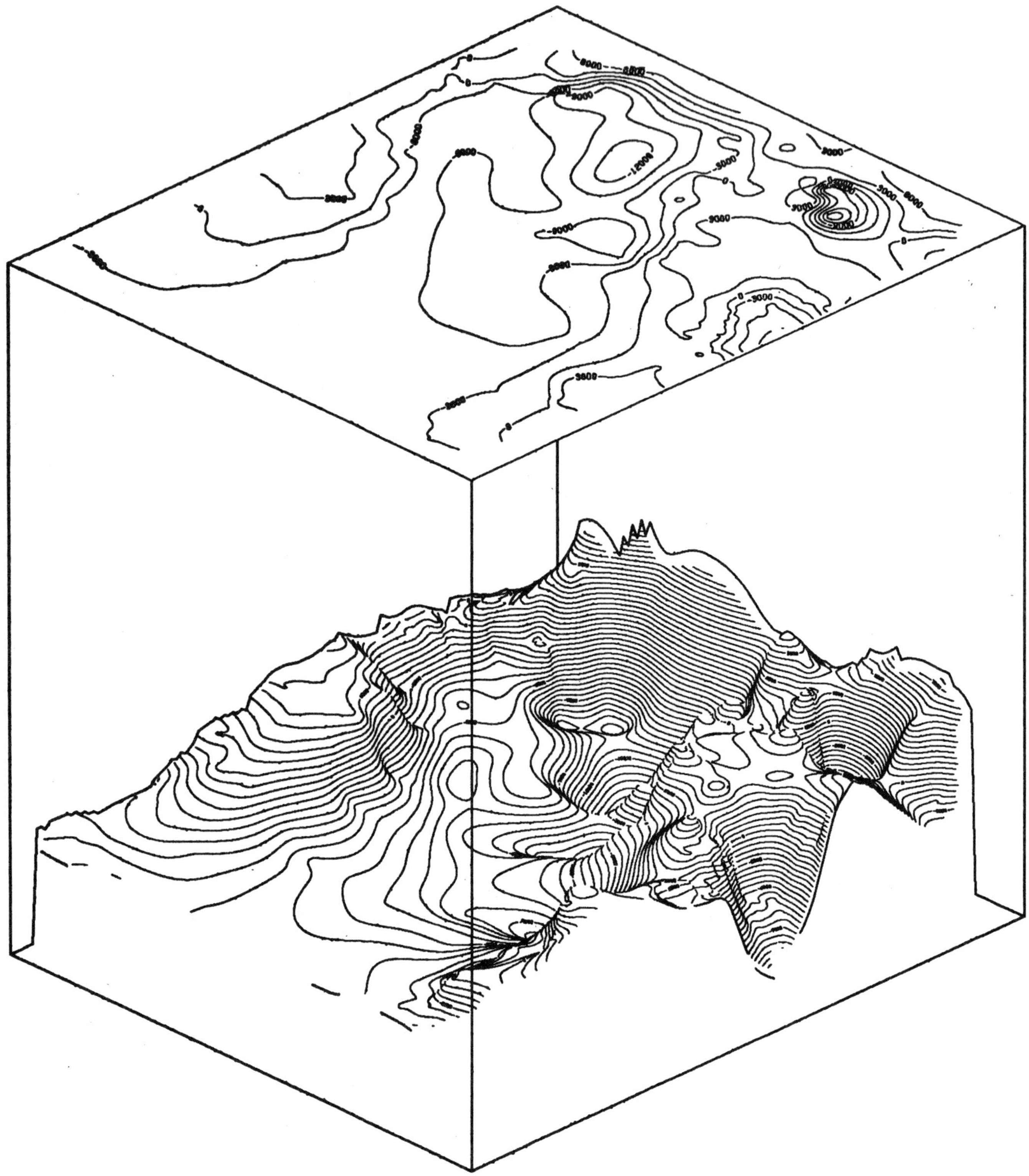

*Figure 5. Raised Contour Display with Ceiling Contours (Mississippian Structure)*

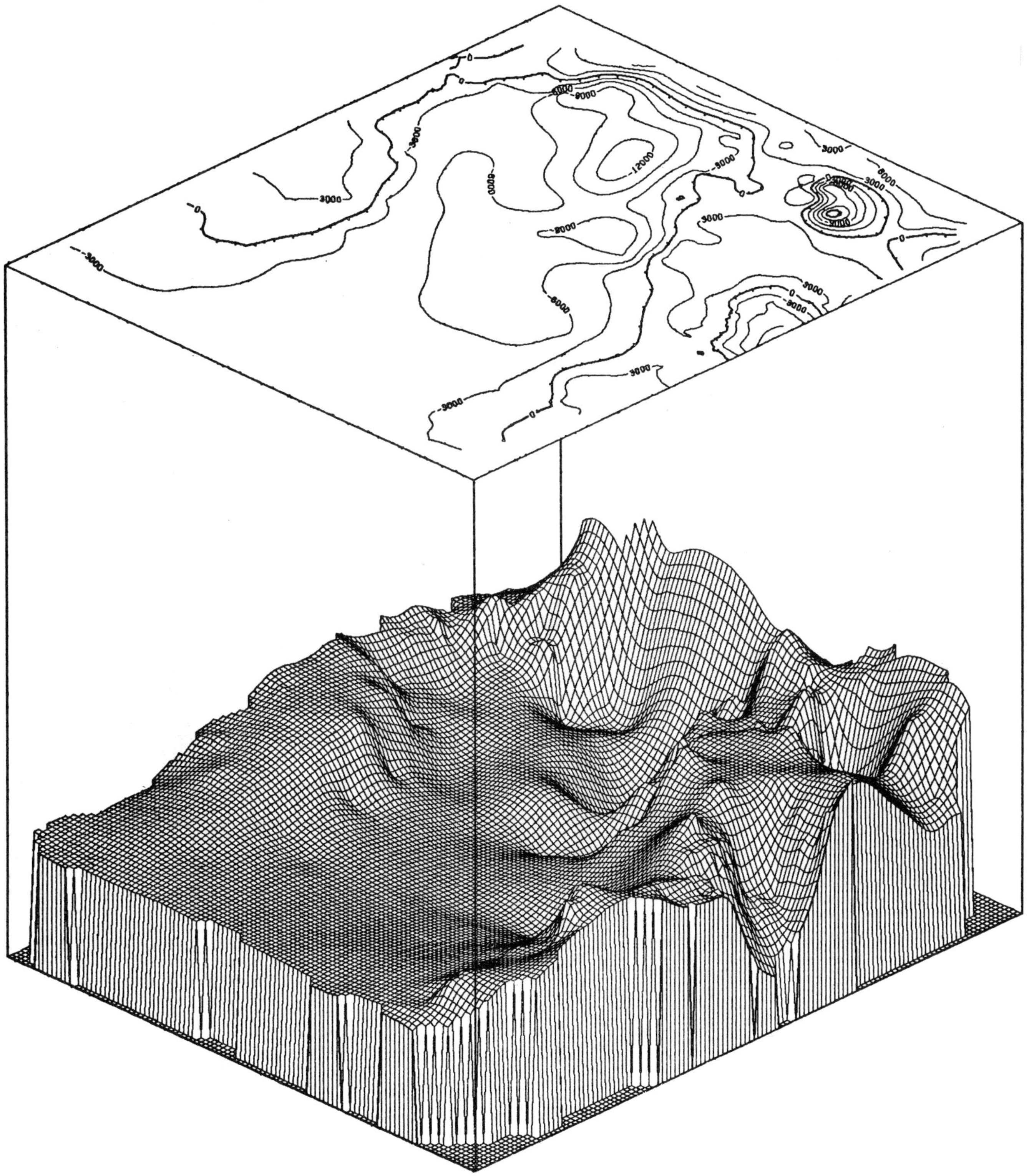

*Figure 6. Isometric Diagram with Ceiling Contours (Mississippian Structure)*

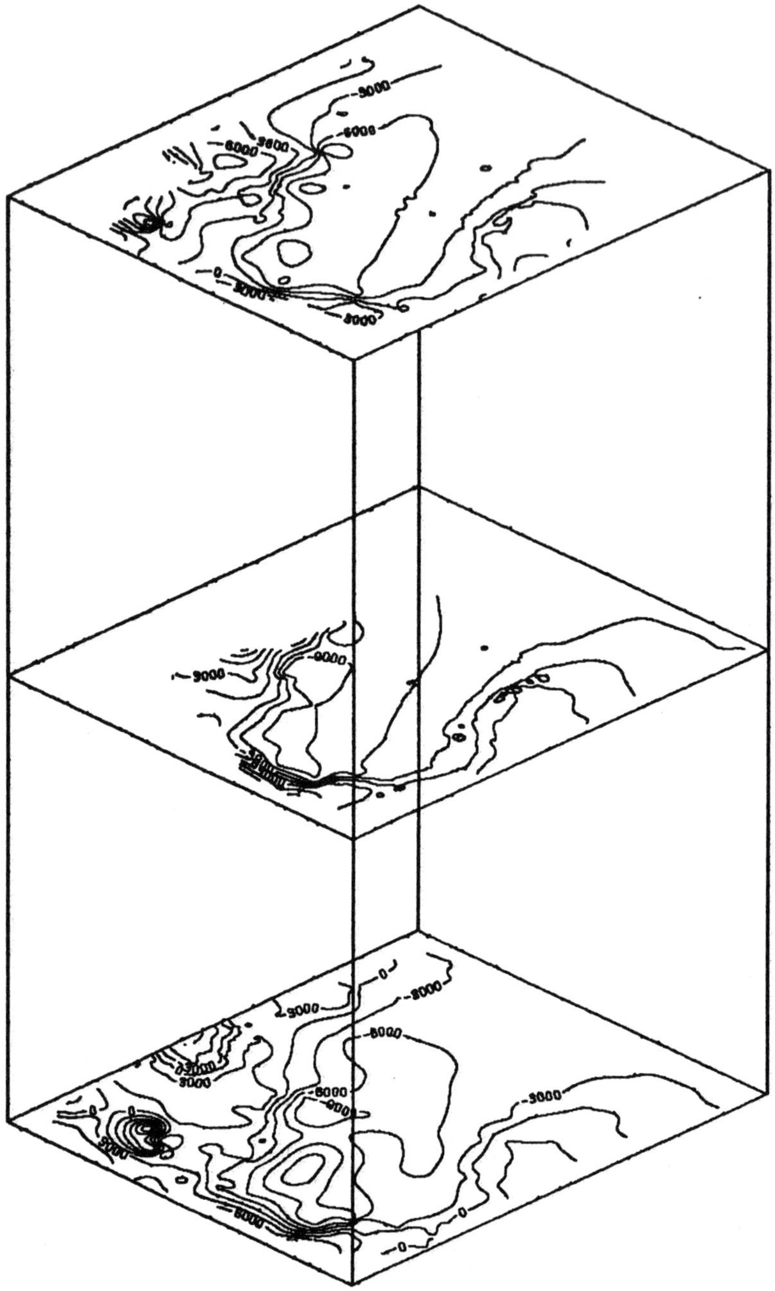

*Figure 7. Stacked Contour Display*

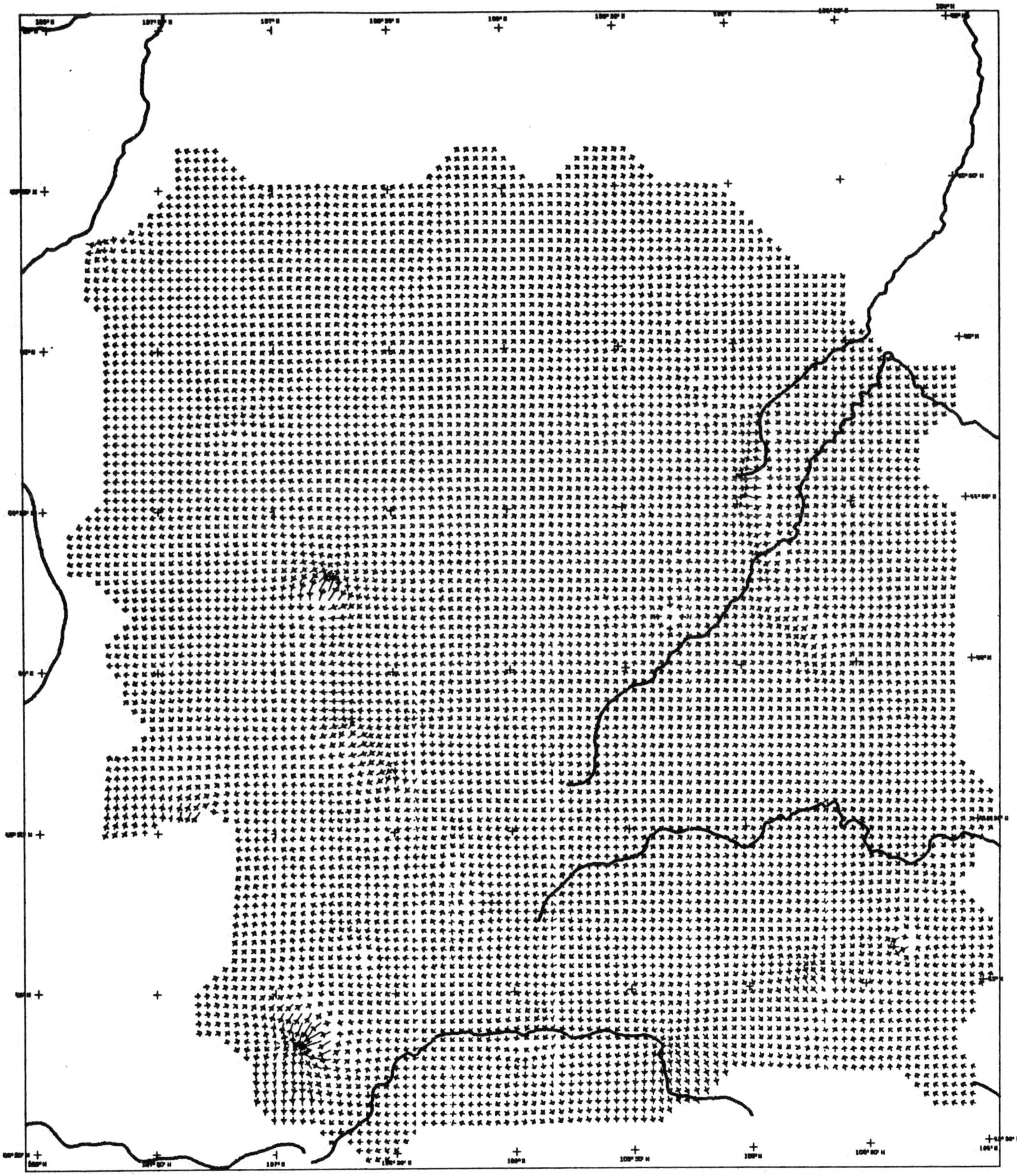

*Figure 8. Gradient Vector Display (Goose EGG Structure)*

# Drill-Hole Data Validation for Subsurface Stratigraphic Modeling

*by Christopher M. Gold*

## THE PROBLEM

Many computer programs have been produced over the years to generate contour maps. While undoubtedly a great deal of duplication has occurred, it is nevertheless true that different types of data and requirements need different solutions in order to generate satisfactory maps. A good example is the recent emergence of digital terrain modeling systems, where it is assumed that data may be acquired from air photographs to any desired resolution by the use of automatic stereo-plotters.

In marked contrast to this is the geological situation where data are acquired at much cost from outcrops or drill holes. Here the mapping problem is the extraction of as much information as possible from the available data, no matter how inaccurate or poorly distributed.

A problem arose in the processing of oil sands drill-hole data for Syncrude Canada. Computer-drawn contour maps of subsurface geological contacts did not bear a sufficiently close resemblance to those hand-drawn by geologists. One cause was incorrect data handling by the mapping program where a contact was absent in a drill-hole. Another was that the contouring system estimated and mapped the top contact of a geological unit even where that unit was absent. In addition, coding errors in the data file were not easy to observe from a computer listing alone, and, finally, hand contouring had permitted a certain amount of data smoothing where the geologist had considered it desirable.

Usually in contouring irregularly distributed data of present or previous topography it is desirable to have the modeled surface pass precisely through each data point. Most techniques, being based on artificial grids generated over the map area, are unable to achieve this. While this is satisfactory for very well sampled smooth surfaces, or for very noisy data such as geochemical soil analyses, topographic data often have both high frequency features and low data point error. An alternative method — subdividing the map area into triangles with a data point at each vertex, and interpolating smoothly within and between the resulting triangles (Gold et al., 1977) constrained the surface to pass through each data point. Both of these approaches were considered somewhat extreme by themselves. It was considered desirable to be able to specify intermediate levels of smoothing, starting with the surface modeling system based on the triangular element data structure of Gold (1977). The project could therefore be considered in two parts. First, an attempt should be made to extend the triangulation-based mapping system to allow for controlled smoothing at data points, and examination of residuals to detect possible sources of error. Second, multiple-surface techniques should be developed so that the partial absence of a geological unit could be handled, and meaningful estimates made in drill holes where the contact under consideration was not observed.

## METHOD DESCRIPTION

### Single-Surface Smoothing, Residuals and Error Detection

Apart from multiple surface comparisons, some error detection procedures may be performed on any resulting single surface. This has been done manually for years, in that data points that were markedly higher or lower than their neighbors showed up clearly during manual contouring and could be verified at that time. Trend surfaces (polynomial surfaces) are also frequently fitted to data points, estimates of the "regional" behavior are made at each data point location and the "residuals" contoured and plotted. Unfortunately polynomials are difficult and expensive to fit to complex surfaces and hence the "residuals" include much of the medium-scale variation as well as any error component. Universal Kriging methods have been used successfully to estimate the "local" component of sample variation, but they are perhaps somewhat constrained by computational cost, acceptance of the methodology, and the need to derive "semivariograms" for each data set.

Given the availability of a triangulation-based data structure (Gold, 1977, Gold et. al., 1977), however, it is possible to make error estimates on a somewhat empirical but more economical and easily understood basis. Each triple of the data points surrounding the data point of interest forms a triangle. This triangular plate can be used to provide an estimate of the elevation at the central point. Thus, in addition to the original value at each data point, a mean and standard deviation may be obtained of the elevation estimates based on the immediate neighbors. In particular, a surface may be deliberately (as opposed to involuntarily) smoothed by replacing each original elevation value by one closer to the mean neighborhood estimate. Figure 1 shows an initial surface with the surface passing precisely through each of the 1,200 data points. Clearly, although this is fairly accurate (the drill-hole elevations are surveyed to within less than a foot), there is a lot of local variation in the elevation of the surface near some data points. Figure 2 shows the resulting surface when the elevations are replaced with the mean values. The major features of the topography are much clearer, yet the deviation of the estimated elevation from its initial value must be much less than the maximum differences between its immediate neighbors, however close or far apart. (No arbitrary fixed smoothing is applied over the whole map area, as is the case when interpolation onto a regular grid is performed.) Figure 3 shows the contoured residuals — clearly, high residuals occur where individual values differ markedly from their relatively consistent neighbors, or else they occur where the neighbors are not in good agreement — as when when a cliff is crossed. It is also clearly possible to adjust or flag only those points where the original elevation value deviates by more than a fixed number of standard deviations from the mean, thus isolating discrepant individual points even more markedly.

### Multiple-Surface Relationships and Error Detection

In order to plot smoothed surfaces or examine residuals for a geological contact based on drill-hole data, it is necessary to be able to handle the situation where one or more contacts in a drill hole were not observed. Figure 4 shows several possible examples. In order to correctly handle missing contacts it is important to realize that geological contacts in sedimentary rocks can be expressed as the top or bottom contact of a geological unit or a combination of units — the top and bottom contacts basically being the topographic surfaces at the end and beginning of sediment deposition, perhaps later modified by erosion. Clearly, maps of these contacts should not be drawn where the unit is absent — this is easy by hand but hard by machine. A unit is absent either because it was never deposited or else was subsequently eroded. These two cases must be handled differently for modeling purposes. In the first case it is assumed that the unit gradually lenses out, and wherever it is first observed to be absent both the top and bottom contacts are considered to have met and to possess the same slope. Where erosion is the cause, it is convenient to consider the upper contact to have been eroded down below the lower contact, giving a negative difference in height between the two surfaces of the unit — for example unit 2 in Figure 4. Thus isopach (or thickness) maps are maps of the positive differences between the top and bottom contacts of a geological unit. This helps avoid the confusion sometimes arising in isopach mapping where the thicknesses are mapped directly and it is not possible to distinguish between true zero thicknesses and "unit absent" information (see Gold and Kilby, 1978). In addition, since it is only legitimate to contour surfaces of units where they are present, plotting should be suppressed where the "difference map" is nonpositive.

Apart from the generation of complete maps, the two-surface model of a geological unit (together with assumptions about erosional or depositional contacts) is needed to help estimate the values of missing drill-hole contacts. On the basis of the preceding discussion, all missing contact values in Figure 4 should be classifiable as not logged, not deposited, or eroded out (in this example the first and third cases only). Interpolated estimates for contacts not logged must then fall between the previous and subsequent contacts, while eroded-out values should be estimated so that the missing unit base contact falls higher than the assumed top contact. Determination of these conditions requires further processing steps.

### Handling Missing Drill-Hole Contacts

Bouille (1976) describes a topological method for digitizing contacts between units on a geological map, obtaining the dual graph, and producing a "summary graph" showing which units are found adjacent to which within the maps area. (A dual graph is one where each zone may be considered to be a "country." A "capital city" is then located within each and "roads" are constructed to connect capitals whose countries have common borders.) Burns (1975) used a similar approach to determine the possible historical sequence of geological units based on older/younger relationships observed at each contact. While the application being discussed here concerns vertical sections rather than horizontal maps, clearly the same principles are applicable.

Figure 4 shows a cross-section of four geological units with drill holes sunk at various locations. The "summary graph" in the lower part of the figure indicates that all units occur adjacent to each other except 2 and 4. Since we know the stratigraphic sequence we can start by considering the 1-2, 2-3, and 3-4 contacts to be depositional: the 1-3 and 1-4, being out of sequence, may be caused either by lensing-out of the units between or by their later erosion. We know from other geological information that they are erosional. We therefore suspect that, since the other contacts involving the base of 1 are erosional, the 1-2 contact also is. This is in fact the case. Figure 4 also shows various ways in which the sequence of contacts observed in the drill holes may not be complete. Any suitable contouring program must be capable of interpreting incomplete drill-hole sequences and arriving at correct interpretations of the reason for the absent contacts.

In practice this is a particularly awkward problem since conventional contact logging may be inconsistent, and hence produce contact sequences with more than one interpretation. This is the case with drill holes B and E, as well as C and F in Figure 4. Some technique is needed to identify these possible conflicts before too much real data are logged. A good method is based on the dual graph concept mentioned earlier. First a "contact naming table" is constructed on the basis of a sketch of the expected stratigraphy. In this table, each overlying unit/underlying unit pair generates a possible contact which should be named. This is frequently trivial since conventionally the top contact of a geological unit is given the same name as the unit; however, where the top contact is erosional rather than depositional (for example the top contact of 3 in drill hole E) this convention is often abandoned.

The second stage requires the construction of the dual table — a unit definition table must be constructed from the contact naming table. It is here that inconsistencies and illegal combinations show up. The table is constructed so that each top/bottom contact pair generates at least one intervening geological unit. Having produced a consistent logging convention for drill holes, and having acquired satisfactory field data, it will then be necessary to generate contour maps of the various stratigraphic units. In the final program it must be recognized that the unit definition table is inadequate in many cases. A

*(Cont'd. on page 57)*

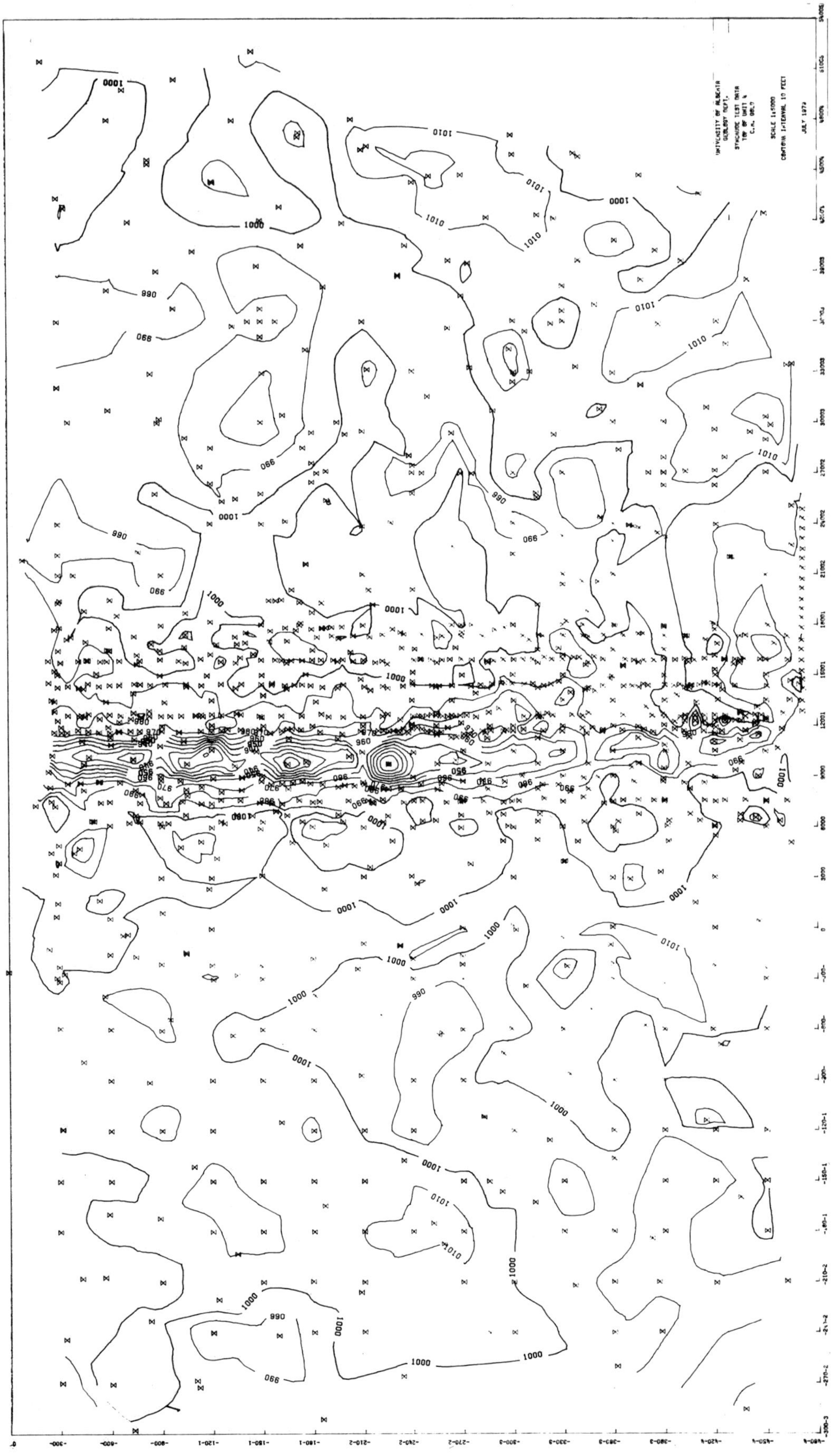

*Figure 1. Contour Map with Surface Passing Through Data Points*

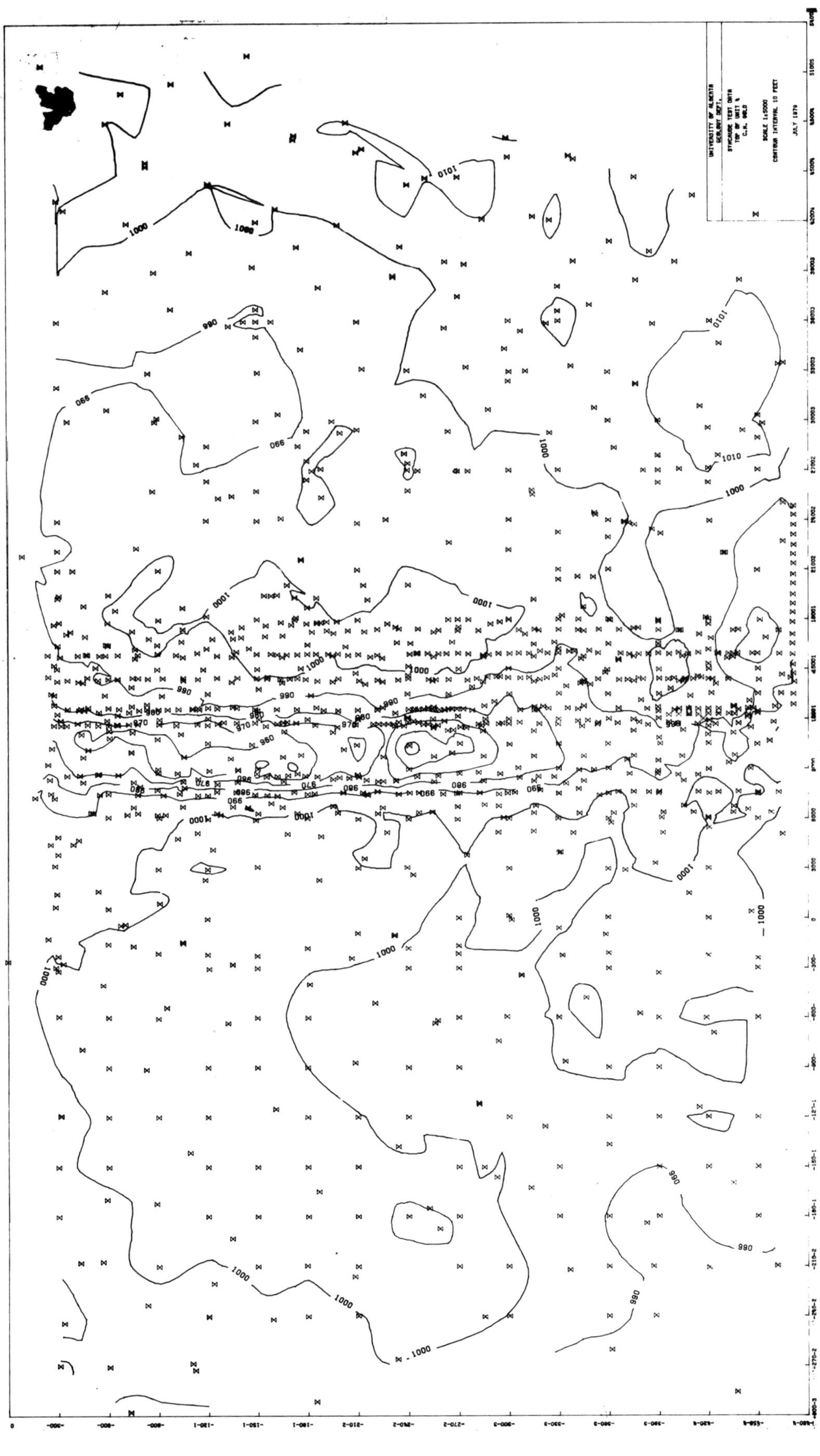

Figure 2. Contour Map of Smoothed Surface

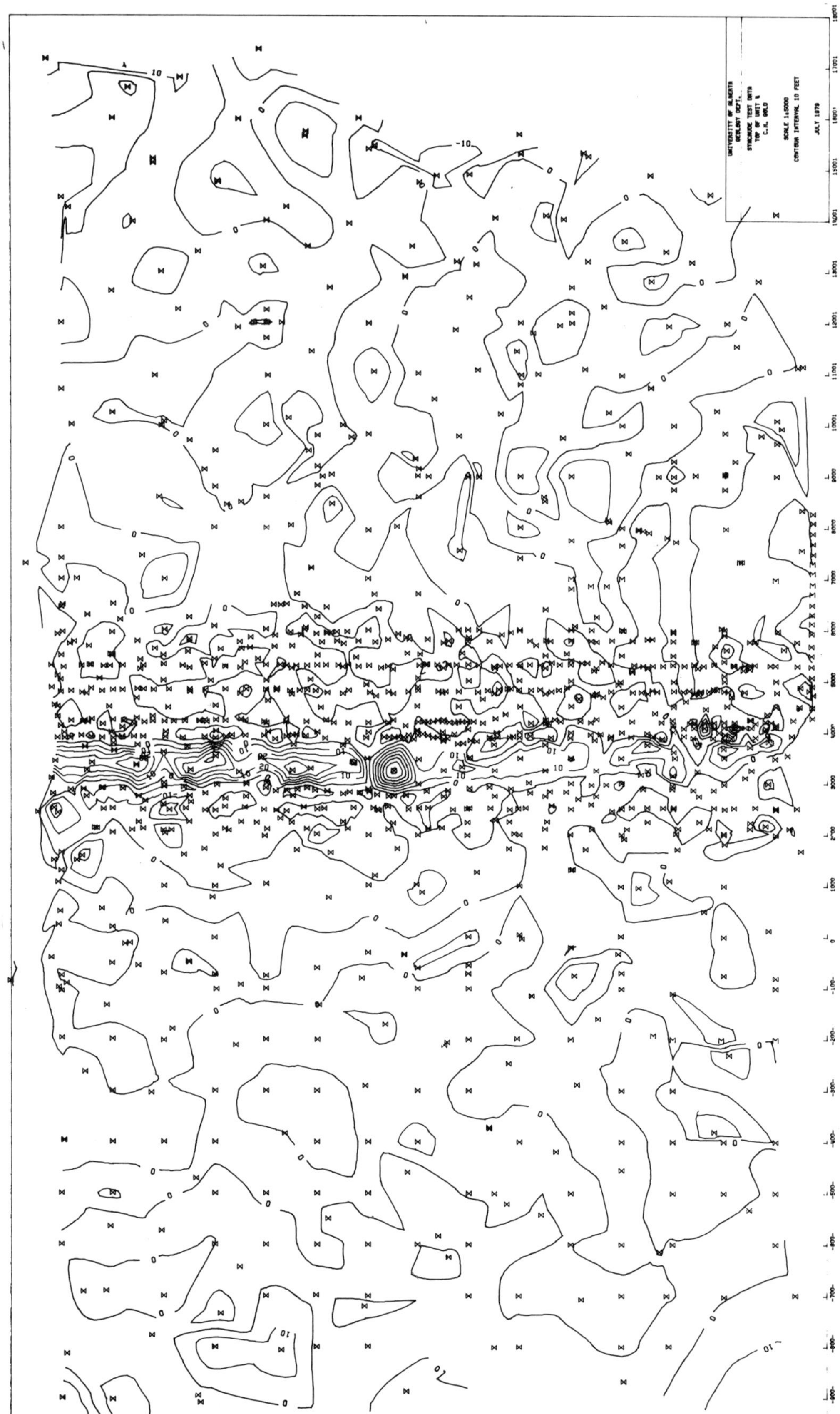


*Figure 3. Map of Residuals (Difference Between Figures 1 and 2)*

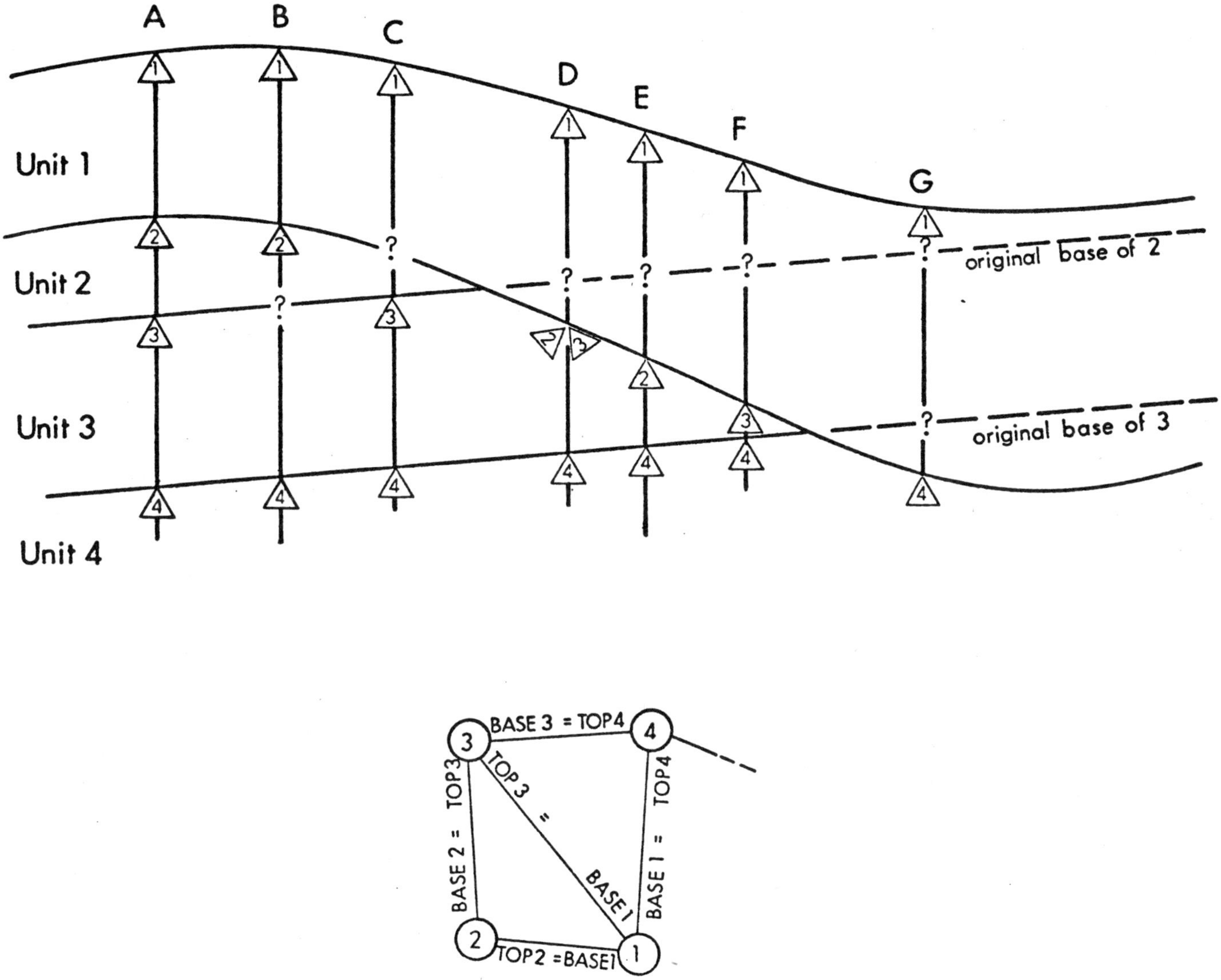


*Figure 4. Sample Topography with Possible Drill-Hole Sequences. Lower Diagram Is Graph of Adjacent Geological Units*

contact-pair observed in a drill hole will affect the specification of the top and bottom of more than one geological unit where intermediate contacts are absent.

## RESOURCES REQUIRED

Let us first summarize the preceding steps. Once the basic stratigraphy is known in the study area, contact naming tables and unit definition tables should be set up to verify the field logging procedure. When an unambiguous set of drill-hole data has been acquired and structure contour and isopach maps are required for various geological units, the data processing program should have the following steps: 1) using the unit definition table as control, separate values should be recorded for the upper and lower contacts of each unit; 2) missing values for each unit should be flagged as absent due to incomplete data logging, absent due to the depositional process, or absent due to erosion; 3) a model should be produced for the top and bottom contact of each unit and missing values interpolated at each drill hole, within the constraints imposed by the drill-hole contacts present and the cause of the lack of data; 4) the structure contour or isopach map should be produced using both top and bottom contacts of the unit simultaneously, either to produce a "difference" map or to suppress plotting where the unit is absent.

The hardware resources required therefore are a medium-size mainframe computer and a drum or flatbed plotter. The software requirements are a mapping package capable of precisely honoring data points, providing local estimates of data point values, handling two surfaces simultaneously and processing drill-hole contact pairs for missing contacts. While it might be possible to add these features to an available contouring program, they are already available using the TRIMAP mapping package developed by the author.

## RESULTS

The above procedure was used to produce maps for Syncrude Canada Ltd, using approximately 1,200 drill holes. Initial maps showed that logging procedures were ambiguous, and hence structure contour and isopach maps did not adequately reflect the data acquired. Syncrude's desire for satisfactory isopach maps, together with structure contours only where the unit was present, promoted the consideration of mapping properties of units rather than contacts. This study promoted the reexamination of ambiguous drill-hole logs, with satisfactory results in the production of the desired contour maps.

## *HINDSIGHT GAINED*

Clearly a problem exists in the processing of drill-hole data. The biggest concern is the use of unambiguous geological contact naming conventions, since without this the data set is to some extent invalid. While this problem occurred with the data set being used, it was possible to correct it without much difficulty. The main aim of the project was to develop procedures to be used for error detection and display of a validly defined data set, and this objective has been met.

## *POTENTIAL FOR TRANSFER*

In the opinion of the author, many drill-hole data sets in use in industry have not been adequately defined, the complications of missing contacts have not been appreciated, nor have the data sets been thoroughly checked for data entry or other errors. Cases are known to the author where geologists avoid using computer displays because the results bear little relationship to the input geology. It is suggested that some of the techniques used here would improve the situation, and it is hoped that the very real problems of defining a computer-usable geological data set will be appreciated by both industry's geologists and its computer specialists.

While the special techniques used are described here for the first time, documentation of the TRIMPA mapping package on which they are based is available from the author.

The author would like to thank Mr. Gerry Lobb and Syncrude Canada Ltd. for their help and advice, and for making the data available. The author was supported in his research by a post-doctoral fellowship from the Alberta Oil Sands Technology and Research Authority.

## *REFERENCES*

Bouillé, F., 1976. "Graph Theory and Digitization of Geological Maps." *Math. Geol.*, v. 8, pp. 375-393.

Burns, K.L., 1975. "Analysis of Geological Events." *Math. Geol.*, v. 7, pp. 295-321.

Gold, C.M., 1977. "The Practical Generation and Use of Geographic Triangular Element Data Structures." *Advanced Study Symposium on Topological Data Structures for Geographical Information Systems Papers.* Laboratory for Computer Graphics and Spatial Analysis, Harvard University, Oct. 1977.

Gold, C.M.; T. D. Charters; and J. Ramsden, 1977. "Automated Contour Mapping using Triangular Element Data Structures and an Interpolant Over Each Triangular Domain." *Computer Graphics*, v. 11, June 1977, pp. 170-175.

Gold, C.M. and W.E. Kilby, 1978. "A Case Study of Coal Resource Evaluation in the Canadian Rockies Using Digital Terrain Models." *Harvard Computer Graphics Week*, July 1978.

# Computer Graphics In Energy Applications

*by Myrle M. Johnson*

## INTRODUCTION

The Energy Systems and Economic Analysis Group and the Statistics Group at the Los Alamos Scientific Laboratory (LASL) are involved in several projects with government agencies requiring energy systems assessment and statistical analysis of energy-related data. Computer graphics provides an efficient and effective tool for displaying, analyzing, and communicating this information.

## HARDWARE AND SOFTWARE

The programs were run on the Control Data Corporation (CDC) 6600 and 7600 computers. The commercial package DISSPLA (Display Integrated Software System and Plotting Language)[2] provided much of the graphics software. The LASL image-processing system, DIADS (Digital Image and Analysis Display System), uses the LADIES (Los Alamos Digital Image Enhancement System) software package, a PDP-11 computer and a COMTAL image processor. Some of the data were incorporated into data-bases using the System 2000 (S2K)[8] hierarchical data-base management system and the GLI-DAMS[7] (Geographical Lease Information Display Analysis and Management System) network data-base management system. Magnetic tapes containing the digitized political boundaries in latitude-longitude coordinates were read to draw the geographic outlines for the maps. Most of the graphic displays were debugged on Tektronix Series 4000 cathode ray tube (CRT) terminals, then processed on an FR80 film recorder as 35 mm slides.

## APPLICATIONS

### Solar Economic Feasibility Studies

The goal of the National Energy Plan (NEP) of April 1977 is to have 2.5 million homes using solar energy by 1985. Economic feasibility analyses are currently underway of the use of solar energy in residential space heating and domestic hot water heating, of alternative energy costs, of various incentive plans, and of sensitivity to heating loads, prices, and potential energy savings. Analyses were first done on a state-by-state basis and are now being refined for some 200 major cities. The polygonal shading technique of the states and counties containing the major cities provides an efficient tool for displaying the hundreds of analyses for easier comparison and comprehension. Several interesting conclusions are being drawn in these studies. "The potential use of solar in residential space heating and domestic hot water applications is measurably enhanced by the proposed incentives. . . . Although the House version of the proposed incentives offers higher total credits, the NEP version is more effective because of the higher dollar credit in the range of most solar system costs (optimal solar fraction). The cost to the federal government (through income tax credits) per barrel of oil equivalent saved would be in the range of one-half to three-quarters the price of foreign supplies."[6]

### Offshore Oil and Gas Leasing Activities

Since 1954 the U.S. Geological Survey has maintained a data-base of lease, production, and revenue information on outer continental shelf oil and gas leasing activities. Computer graphics has been used to describe and compare the data by company, lease, and sale.[9] The graphic displays include multi-axis graphs for time profiles of leasing and production activities, Chernoff faces, and truck representations of company data and color-coded correlation matrices. Computer-generated lease maps are also being made where the individual leases are delineated by survey lines. Statistical analyses of these data involve fitting probability distributions to the bids, studying joint/solo bidding practices among the oil and gas companies, determining the relationship of government pre-sale values and production, performing a discriminant analysis of producing versus nonproducing leases. In working with such large data-bases, it is significant to note that computer graphics serves not only as an effective analytic and communication tool, but also as an efficient means of checking both the quantity and the quality of the data stored. It is upon such data-bases that good analyses are founded.

### Electric Power Production Impact Studies

Under the auspices of the Department of Energy (DOE) and the Environmental Protection Agency, the environmental, social, and economic impacts of electrical power generation by utilities belonging to the Western Systems Coordinating Council (WSCC) are being studied.[3] They are concerned about the boom town problem, the geographic distribution of changes in the death rate, the changes in life expectancy, the public health effects and costs, and the pollution dispersion patterns due to the siting of power plants.

Still another concern is the effect of various contaminants from these power plants on scenic landscapes.[5] The Clean Air Act of 1977 places special emphasis on the protection of national parks and wilderness areas. Computer-simulated images aid in this study. Photographs of landscapes are scanned on a microdensitometer and digitized for wavelengths corresponding to red, blue, and green light. Data on transmitted and total radiance from plume particles emitted from power plants are calculated and converted to equivalent film densities by a plume visibility code. Image-processing techniques are used to superimpose these densities onto the image to form the smoke plume. The color and extent of the cloud are determined by the type of pollutant, different control technologies and varying meteorological conditions.

### National Uranium Resource Evaluation

The objective of the National Uranium Resource Evaluation Project's nationwide airborne and stream sediment reconnaissance surveys is to classify regions with respect to their potential mineralization.[1] Graphical and analytical techniques used here include scattergrams, 3-D density plots, color intensity plots, a linear discriminant analysis display, contour maps, and overlay maps of sample locations and mineral concentrations. The massive amounts of data associated with this project are attributable to recent advances in chemical and geological data collection technology. Hence, data-base management systems and innovative computer graphics techniques are required to extract information from the data.

### Regional Energy Studies

To address our nation's energy-related problems, it is important to gain regional perspectives on the environmental, socioeconomic, legal, and resource constraints on energy production and consumption and to assess the regional implications of national energy planning. Techniques such as composite mapping and flow mapping help to give this perspective.

The General Map Analyses Planning System (GMAPS)[10] is an interactive geo-information system with the capability for composite mapping. Applications of this system to date include regional energy development studies (Regional Studies Program and National Coal Utilization Assessment for the DOE), national park land use planning (National Park Service), and impacts on fish and wildlife habitats (U.S. Fish and Wildlife Service). Baseline maps of locations of natural resources, transportation corridors, wilderness areas, and legal, social, and environmental constraints for selected regions in the U.S. are stored in the computer in matrix format. Composite or overlay maps are created from two or more baseline maps using the arithmetic (numeric) and logical combination modes. Numeric combination allows the use of the addition, subtraction, multiplication, and division operations on weighted baseline maps. Logical combination allows the 'and,' 'or,' and 'not' operators to merge maps on a cell by cell basis.

Flow maps provide a technique for summary overviews of the energy production-consumption-export system used in the DOE Regional Energy Assessment Program. Such energy flow displays show regional production and exports by type of natural resource (hydroelectric, wood, coal, natural gas, oil and natural gas, uranium) and regional consumption by sector (residential, commercial, industrial, transportation).[4] Other techniques use histograms and pie charts superimposed on maps to communicate the regional energy scenario. Displays include regional, state, and county political boundaries as well as air quality control regions and business economic areas (BEA).

### Energy Resources Studies

In addressing our nation's energy-related problems it is also important to assess world energy resources. Data on population, production, consumption, exports, and imports for various types of natural resources are displayed by means of histograms, pie charts, flow maps and polygonal shading of countries on a world map.

## *SUMMARY*

Applications of computer graphics techniques to energy-related studies have been shown to be a valuable tool in the display, analysis, and communication phases of these studies.

## *ACKNOWLEDGEMENTS*

I wish to thank the members of the Energy Systems and Assessment Division for permission to use their slides and for the briefings on their respective projects. The authors listed in the References can be contacted for more detailed information.

## *REFERENCES*

[1]Bement, T.R.; D. V. Susco; D. E. Whiteman; and R. K. Zeigler. "National Uranium Resource Evaluation Program." Los Alamos Scientific Laboratory Report LA-6804-PR (May 1977).

[2]DISSPLA, Integrated Software Systems Corporation, P. O. Box 9906, San Diego, CA.

[3]Ford, A. and H. W. Lorber. "Methodology for the Analysis of the Impacts of Electric Power Production in the West." Los Alamos Scientific Laboratory Report LA-6720-PR (Jan. 1977).

[4]Kolstad, C. D. "The 1975 Energy Production System in the States of the Rocky Mountain Region." Los Alamos Scientific Laboratory Report LA-6624 (Jan 1976).

[5]Leonard, E. M.; M. D. Williams; and J. P. Mutschlecner. "The Visibility Issue in the Rocky Mountain West." Los Alamos Scientific Laboratory Report LA-UR-77-2558.

[6]Roach, F.; S. Noll; S. Ben-David; L. Bickle; and W. Schulze. "Prospects for Solar Energy: The Impact of the National Energy Plan." Los Alamos Scientific Laboratory Report LA-7064-MS (Dec. 1977).

[7]Sibert, J. L.; R. Phillips; and J. Lohrenz. "A Data-Base Management System for Spatial Display of Federal Offshore Oil and Gas Lease Data." Los Alamos Scientific Laboratory Report LA-UR-77-2031, presented at the Second Geological Survey Computer Symposium, Reston, VA (Mar. 1978).

[8]System 2000, Data-Base Management System, MRI Systems Corporation, P.O. Box 9968, Austin, TX.

[9]Waller, R. A.; E. A. Monash; and J. Lohrenz. "Some Computerized Graphic Technical Applications for Federal Mineral Lease Management Support." Los Alamos Scientific Laboratory Paper LA-UR-77-583, presented at the First Computer Symposium, Reston, VA (Mar. 1977).

[10]Wecksung, M. J.; R. A. Wiley; and A. K. Turner. "GMAPS User's Manual." Los Alamos Scientific Laboratory Report LA-6975-M (Jan. 1978).

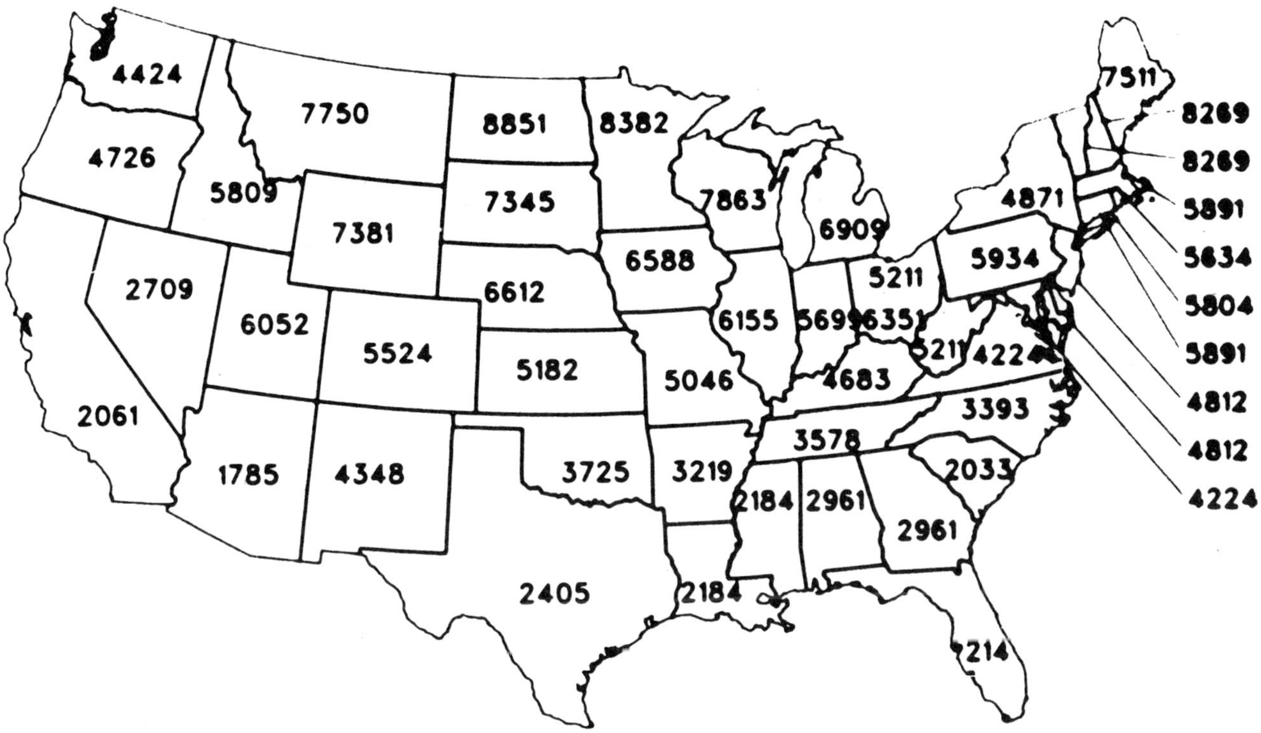


*Figure 1. Solar economic feasibility studies. Heating degree days by a city representative of the state*

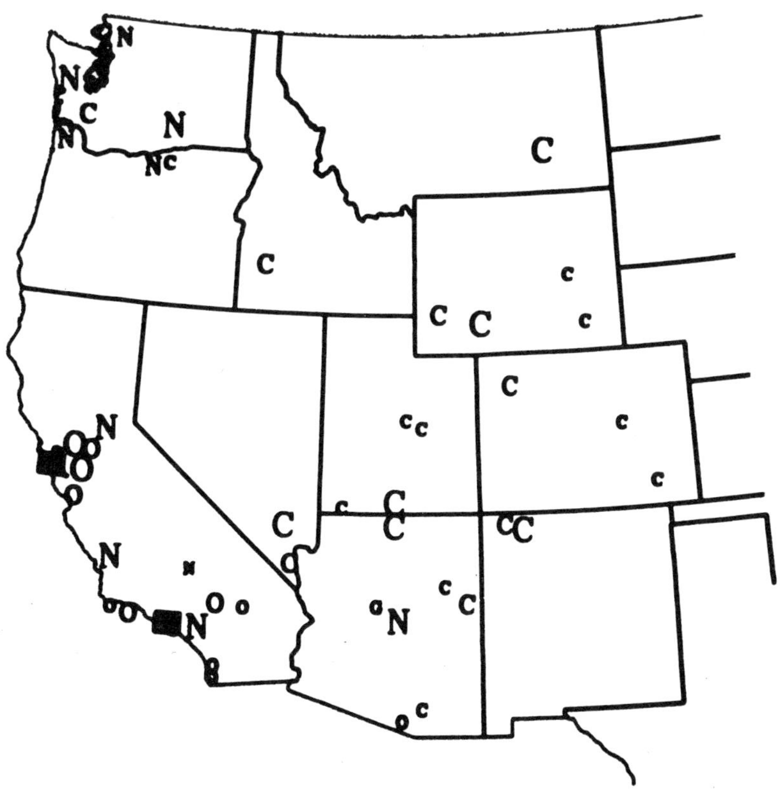

*Figure 2. Electric power production impact studies. Proposed and existing WSCC power plants by type and size*

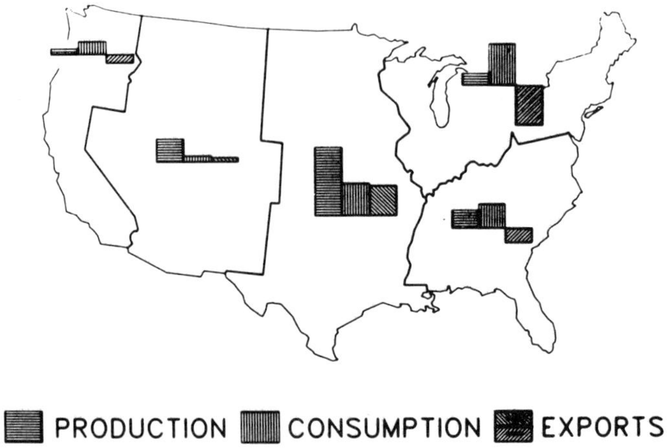


*Figure 3a. 1975 Regional Energy totals*

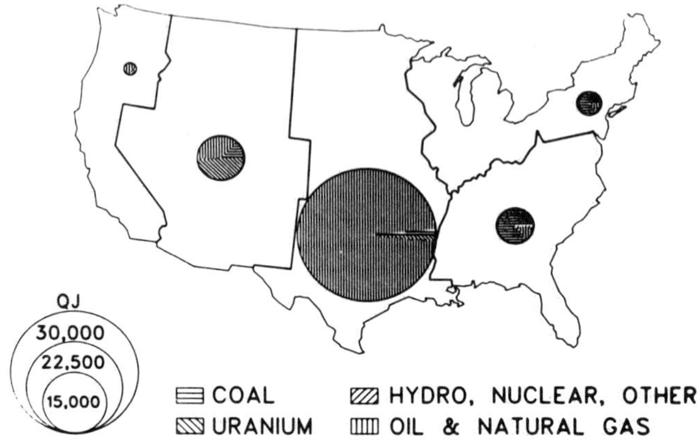


*Figure 3b. 1975 Regional Energy Production*

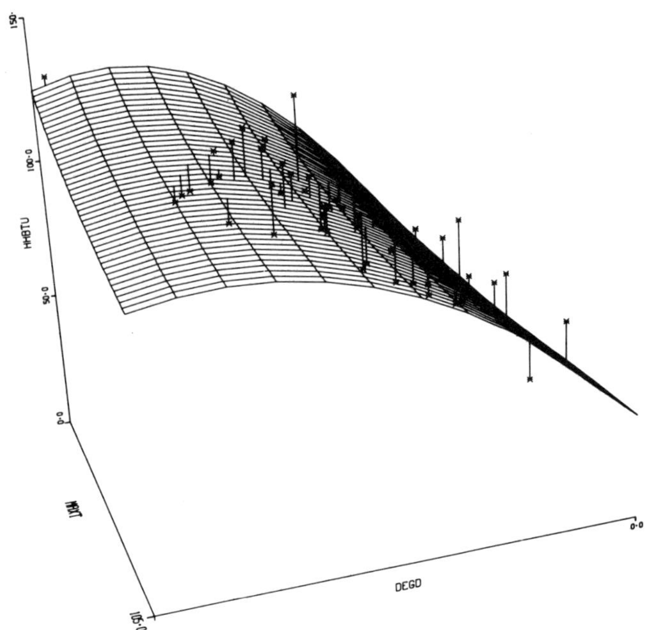


*Figure 3c. 3-D display of a linear regression analysis for household BTU consumption per capita for the fifty states and D.C.*

*Figure 3. 3 examples for energy studies*

# The Council on Environmental Quality Air Pollution-Health Risk Model

*by Kay H. Jones, Randi L. Ferrari, and Timothy L. Chapman*


## *ABSTRACT*

An air pollution-health risk model for relating ambient air quality data, population statistics, and health impacts has been developed by the Council on Environmental Quality. The model was designed to quantify the health risks of air pollution in metropolitan areas. The concept and mechanics of the model were demonstrated by an analysis of the relationship between CO exposure and the related increase in angina attacks for stable angina pectoris patients. This analysis demonstrated that the risk of CO exposure to stable angina pectoris patients appears to be quite low, although the authors have not necessarily concluded that the risk is low for other cardiovascular disease categories. Rather, they are advocating a method of analysis which provides greater insight into the actual risk associated with various levels of pollution. Such modeling results should facilitate sound regulatory judgments.


## *INTRODUCTION*

CEQ has developed an air pollution-health risk model for relating ambient air quality data, population statistics, and impacts on health. This model is designed to quantify the health risks of various air pollutants in metropolitan areas. The flexibility of the model enables the analyst to calculate population exposure risk for any pollutant for which there are air quality data. Specific health risks can be estimated if a damage function can be approximated. Variables such as mobility and indoor/outdoor factors can also be added to the model.

The concept and mechanics of this model are demonstrated by an analysis of the relationship between CO exposure and the related increase in angina attacks for stable angina pectoris patients. Angina pectoris is a heart condition in which chest pains are precipitated by deficient oxygenation of the heart muscle. Stable angina pectoris patients exhibit clinical angina symptoms but no recognized myocardial infarction. Risk is quantified by estimating the annual number of excess angina attacks resulting from CO exposure.

This model can give public policy-makers a clearer picture of actual health risks in the most populous and polluted cities and provide a visual and quantitative representation of present and future population exposure risk.

Analysis of exposure risk will help focus health research on pollutant concentrations and frequencies similar to those found in the environment. This approach is more useful than analyzing the effects of exposure to extreme levels of pollutants and then attempting to extrapolate downward to ambient conditions. Studies which lend themselves to the generation of continuous damage curves should emerge. The air pollution-health risk model may also illustrate air quality monitoring deficiencies.

## *THE AIR POLLUTION-HEALTH RISK MODEL*

The air pollution-health risk model, illustrated in Figure 1, can be used to analyze either one pollutant and its one or more known effects, or an aggregate of pollutants and their effects. To simplify discussion, the authors have analyzed one pollutant relative to one effect. Several iterations of this approach will produce an aggregate list of risk data for a single pollutant and its perceived damages.

It is difficult to quantify directly the *in situ* health risk experienced by a population which may be attributable to a single pollutant. However, one can extrapolate from clinical studies of single pollutant effects on subsets of the population. One can also extrapolate from epidemiological studies of site-specific populations to estimate the health risk for the entire population. The functional relationship between the health risk and a particular exposure to a given pollution concentration is called a damage function. The most desirable form of this function is one which is continuous; however, step functions will work. Most continuous functions are assumed to be linear because of the uncertainties in their derivation. Once the damage function is approximated, it can be integrated with ambient exposure information.

One can calculate risk rates for each monitoring station by integrating the frequency distribution of pollutant concentrations (the number of days per year at each concentration, hereafter referred to as the frequency distribution) for each monitoring site with the damage function. Integration is performed by first multiplying the risk constant (the slope of the damage function) by each pollutant concentration interval (above any predetermined threshold) and then summing. In this study it has been assumed that health risk to pollutant lev-

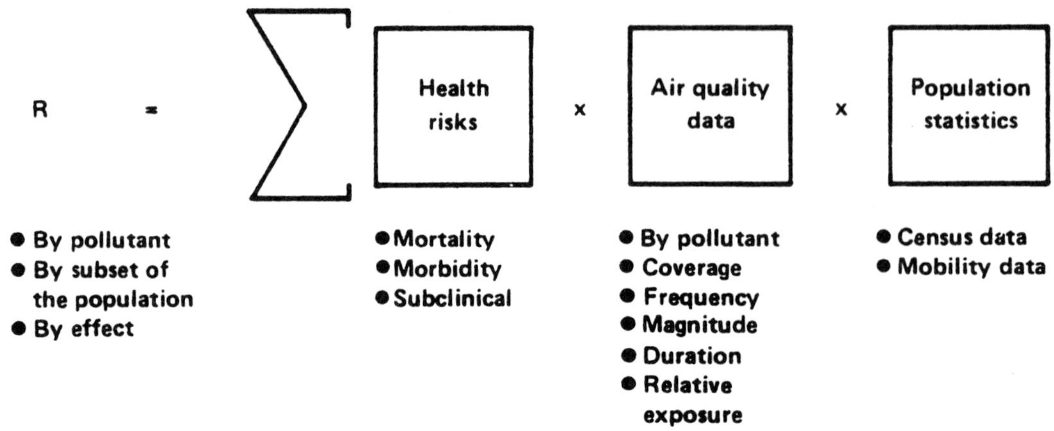


*Figure 1. CEQ Air Pollution-Health Risk Model*

els below the National Ambient Air Quality Standards (NAAQS) are either highly uncertain or insigificant and therefore the authors show the health damage function beginning at the standard (Figure 2). This assumption is valid if the standard-setting protocol established by the Environmental Protection Agency is followed in a rigorous fashion. Standards are to be set with a reasonable margin of safety below levels which are known to affect the more sensitive segments of the total population. The risk to the sensitive population must then be assumed to be of very low probability at such a standard and hence the health risk below such a standard cannot be rationally estimated.

The frequency distribution of ambient pollution levels probably should be corrected because such measurements do not represent the actual population exposure to air pollution. Variables such as monitoring site configuration, indoor/outdoor relationships, and near source/far source factors must be considered. Despite all these confounding factors, fixed monitoring station data can be used as a surrogate for population exposure if the influence of these factors on actual exposure can be reasonably accounted for. The indoor/outdoor and near source/far source relationships have been examined in a separate paper by the authors.[1]

The public health risk of ambient air pollution is generally a problem of urban areas because population densities are high and most pollution occurs there. Therefore an appropriate unit to analyze is the Standard Metropolitan Statistical Area (SMSA). Population statistics are available for SMSAs at the census tract level. The total population at risk within the SMSA is derived in the following way: the risk rates are integrated with (simply multiplied by) the population data at the census tract level and then summed over all census tracts. Subsets of the total population, representing those individuals susceptible to the health effect being evaluated, can be used in lieu of the total census tract population. For example, these subsets can isolate age, sex, race, income, and disease categories or combinations thereof.

Air pollution levels in central business districts tend to be quite different from those in the surrounding suburban areas. For example, high CO levels tend to be localized, occurring in downtown street canyons with high traffic volumes; the transport and transformation of hydrocarbons tend to cause higher levels of ozone in suburban areas downwind of the urban centers which produced the high hydrocarbon emissions. Therefore mobility of the poulation should be considered when estimating population exposure. In order to estimate the change in population exposure that results from the influx of people into the center city during the day, two sets of population figures can be derived: census tract resident population (Census Bureau Data) and census tract population during working hours derived by subtracting the census tract residents employed from the resident population and adding the number of workers in the census tract.

The Census Bureau's 1970 Urban Transportation Planning Package analyzed the number of workers in census tracts, but only a small percentage of the population in a few urban areas was sampled. Better data will be obtained in the 1980 census. If population exposure occurs during working hours, the risk rates should be integrated with "at work" population figures. Population exposure occurring throughout the day may require a segmented approach in which different risk rates are used for the same population but at various times and locations, e.g., four hours at work downtown, one hour commuting, and three hours at home.

## *THE CO HEALTH RISK MODEL*

Carbon monoxide is emitted almost exclusively by cars and other gasoline-powered vehicles. Variation of CO levels is caused by changes in the emission rates and traffic densities normally associated with the morning and evening rush hours. CO is a health hazard because it combines with the oxygen-carrying hemoglobin in the blood more easily than oxygen, forming carboxyhemoglobin (COHb) and thus constraining oxygen transport. The air pollution health-risk model is used to access CO exposure effects on stable angina pectoris patients. The model quantifies the health risk by estimating the annual number of excess angina attacks associated with daily CO exposure. It is estimated that 2.1 percent of the national population suffers from stable angina pectoris. Most stable angina patients suffer from one to five involuntary attacks per day from all causal factors.[2]

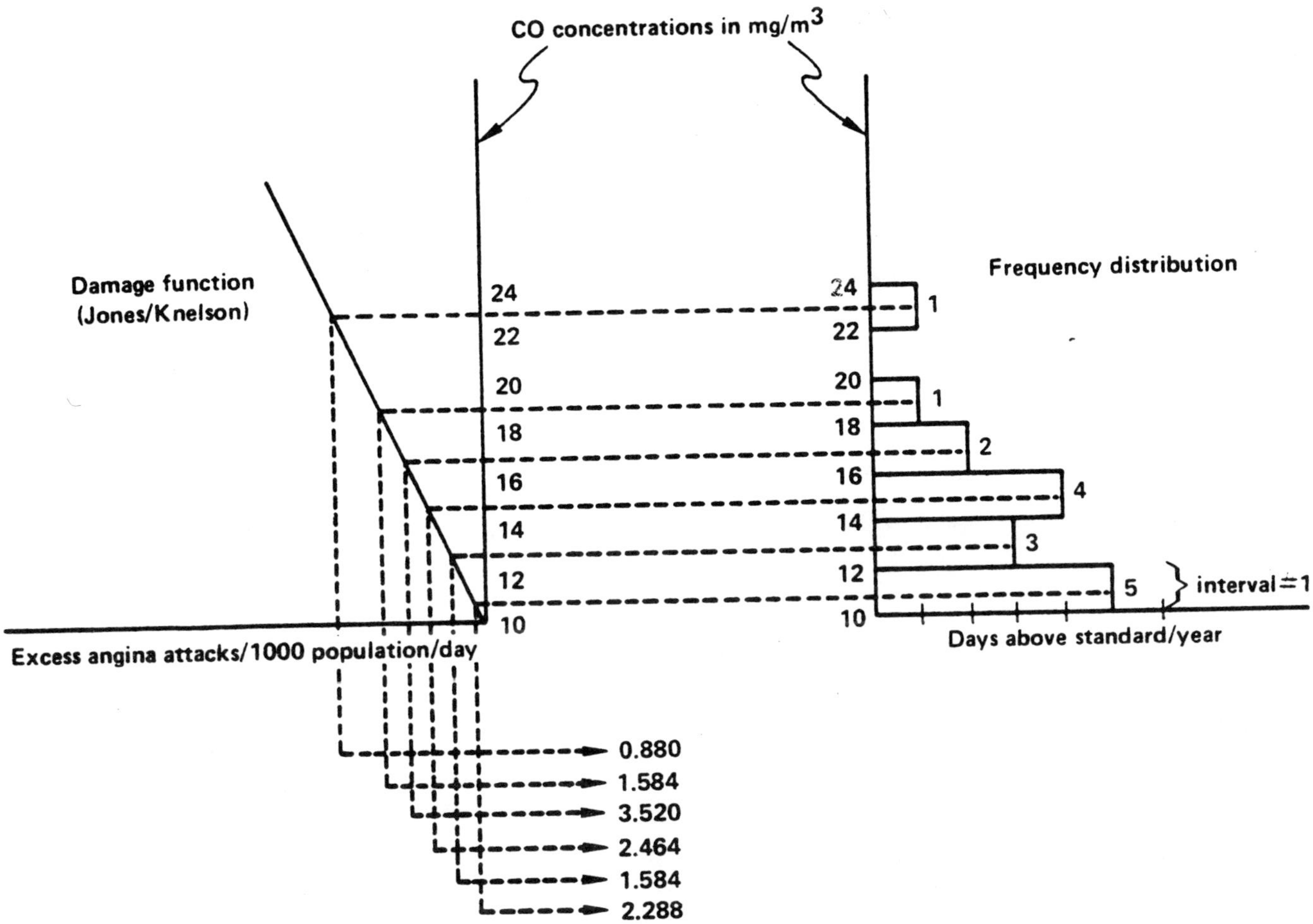


**Risk rate for sample station = Σ = 12.320 Excess angina attacks/1000 population/year**

**- - - ➜ Dotted arrows signify integration process**

**Formula used:** $0.176 \times \left( \begin{array}{c} \text{concentration in mg/m}^3 \\ \text{of interval midpoint} \end{array} - \begin{array}{c} \text{CO standard in mg/m}^3 \\ \text{[or the Y intercept]} \end{array} \right) \times \text{frequency} = \text{Risk rate}$

**For example, interval #1: 0.176 x (11–10) x 5 = 0.880**

*Figure 2. Methodology for Risk Calculations*

## METHODOLOGY

The 1975 CO data used in this study were from UPGRADE's Pollutant Standards Index (PSI) data-base, which contains daily maximum eight-hour average values retrieved from the Environmental Protection Agency National Aerometric Data Bank. Hard-copy summary data for the entire nation were reviewed to locate those SMSAs which had at least three stations showing violations of the national ambient CO standard. The SMSAs were screened in this manner because it was assumed that in the mapping routine three data points were needed to get spatial resolution. In this review, 26 SMSAs qualified. The data were then screened to eliminate any stations that monitored less than aproximately half the year and/or had fewer than 100 days of monitoring. After this screening 20 SMSAs had at least three stations with violations (Table I); nine SMSAs were analyzed for this report. Philadelphia had the highest number of stations monitoring CO but had the worst yearly coverage of the 20 SMSAs that met the screening criteria.

The CO risk rates were derived from integration of the frequency distributions of days above the NAAQS for CO (10 mg/m$^3$) produced by the UPGRADE system with the damage function developed by Jones and Knelson, as shown in Figure 2.[3] The Jones-Knelson damage function illustrations the relations between eight-hour CO exposure and the number of additional angina attacks expected per 1,000 population. The Jones-Knelson function is based on three clinical studies[4 5 6] which related the decreased time between angina attacks to CO exposure. This decreased time was translated into risk of additional attacks. Because the relationship between blood carboxyhemoglobin levels and CO exposure is known, the risk

Table I.
SMSAs with Three or More Stations Having Violations and Sufficiently Monitoring CO in 1975

| SMSA Name | # of Stations | # of Stations Having Violations |
|---|---|---|
| *Boston | 6 | 4 |
| Chicago | 12 | 7 |
| Dayton | 3 | 3 |
| *Denver | 6 | 6 |
| Detroit | 5 | 3 |
| Los Angeles | 13 | 13 |
| *Louisville | 9 | 9 |
| Milwaukee | 9 | 4 |
| Minneapolis | 4 | 3 |
| Newark | 4 | 4 |
| New York | 16 | 10 |
| Oakland | 10 | 4 |
| *Philadelphia | 17 | 12 |
| *Phoenix | 8 | 7 |
| *Portland | 5 | 5 |
| Riverside | 14 | 8 |
| *St. Louis | 9 | 8 |
| *Seattle | 4 | 4 |
| *Washington, D.C. | 11 | 9 |
| Wichita | 4 | 4 |

*SMSAs included in this study

of additional attacks can be equated to eight-hour CO exposure using the formula: % COHb = 0.144 x mg/m$^3$ CO.[7] The number of angina attacks per 1000 population per day is then calculated by subtracting the standard (10 mg/m$^3$) from the CO levels above the standard and then multiplying by 0.176. It is assumed that an eight-hour daily maximum value will achieve saturation of COHb in the blood. Saturation means that a maximum level of COHb has been reached for the level of CO exposure. Any further exposure to that CO level will not increase the COHb in the blood.

The frequency distribution and the risk function are integrated by taking the midpoint of each interval, subtracting 10, multiplying by 0.176, and summing (Figure 2). This calculation produces a risk rate in attacks per 1000 population per year for the station analyzed. As discussed previously, any stations with all values below 10 mg/m$^3$, the national standard, were assigned risk rates of zero. In this analysis, the frequency distributions were not adjusted for indoor/outdoor variation. It was assumed that the entire population was being exposed to monitored levels. This yields conservative estimates (see *Refinements*).

The risk rates for each station are plotted by the Harvard Synagraphic Mapping System (SYMAP) and isopleth lines of risk rates interpolated. Boundary conditions were set by placement of dummy station risk values of zero approximately 20 miles out from the center of each city. It is reasonable to assume that violations of the eight-hour CO standard do not occur at such a distance from most urban centers.

In this report the risk data are visually presented by an overlay of the isopleth lines of risk on population density maps from the Census Bureau's *Urban Atlas*. The data could also be presented by plotting the total number of excess angina attacks calculated for each census tract. These figures are calculated by multiplying the population census tract data by the interpolated risk rates within the census tracts.

## RESULTS

The isopleth line of risk rates are overlaid on the population density map in Figure 3 for Denver. The numbers on the isopleth lines represent the annual number of excess angina attacks experienced per 1000 population as a result of CO exposure. The number in the center of the isopleths is the highest calculated risk rate in the SMSA. For illustration purposes, only part of the SMSA was used because risk occurred primarily near the center of the city. The census tracts surrounding the center city area had much lower population, low or no CO violations, and hence negligible or no angina risk as a result of CO exposure.

There are several ways to analyze these data. One can analyze the risk experienced by individual patients in reference to their locations, or one can analyze the total risk experienced by stable angina pectoris patients resulting from their CO exposure in the SMSA. Cardiologists try to minimize the number of attacks their patients experience through reduction of stress and exercise and/or drug therapy.[8] Risk is therefore best expressed in terms of excess attacks per patient. The annual attack rate by SMSA allows one to compute the average daily risk of attack per angina patient. Further, the site specific risk numbers provide for calculating average daily and maximum daily risk by location within each SMSA. These data for the nine SMSAs studied are shown in Table II, where the average

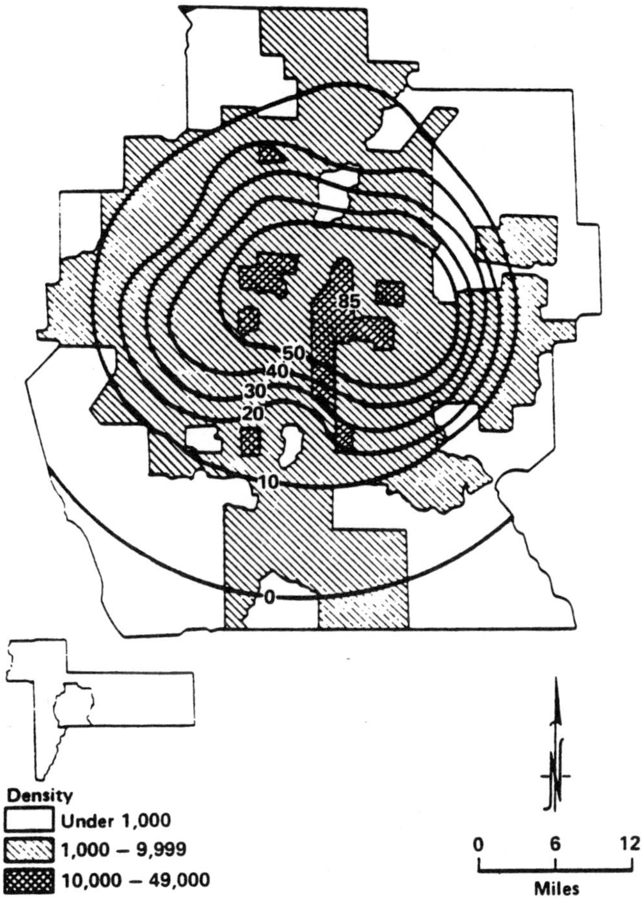


*Figure 3. CO-Related Angina Attacks, Denver 1975*

daily and maximum daily risks by location are calculated only for the area having the highest risk within each SMSA. These rates are conservative because they are related to outdoor CO measurements which are near the source. It appears that the SMSA average daily risk is quite low, 0.0002 to 0.0037 attacks per patient. The average daily per patient risk in the highest exposure areas of each SMSA appears low as well, 0.0011 to 0.0170 attacks. The day with the highest level of CO at the site that most frequently violates the standard represents a maximum risk of approximately 0.025 to 0.260 attacks per patient for that one day. Recall that all these risks are for the base year 1975.

In terms of relative risk, it does not appear that CO exposure contributes significantly to the angina attack risk attributable to all causes. If we assume that the average stable angina patient currently suffers from one to five attacks per day, CO exposure appears to contribute a maximum of 0.122 percent of the total angina attack risk in the worst SMSA analyzed (Table III). SMSAs other than Denver have estimated CO risk levels of 0.007 to 0.009 percent of the total angina attack risk. Denver's CO risk level is 0.122 percent of the total attack risk, causing 34.1 thousand attacks. Although this analysis has yet to be done for Los Angeles and New York City, one would expect their rates to be similar to that in Denver because of similar air quality profiles.

The predicted risk in Denver, assuming the expected 50 percent reduction in emissions by 1985 and using a proportional rollback model, is illustrated in Figure 4. Because the reduction in violations is disproportional to the reduction in emissions,[9] an 85 percent decrease in the number of predicted excess angina attacks per year as a result of CO exposure is forecast, which calculates to 5.1 thousand attacks. This statistical phenomenon is illustrated in Figure 5, in which the areas under the curves and to the right of the air quality standard line represent the number of violations in days above the standard. The upper end of the predicted distribution, the maximum daily concentration, is 50 percent less than the present maximum daily concentration, and the shaded area of the predicted distribution is 85 percent less than the present distribution.

## *REFINEMENTS OF THE CO HEALTH RISK MODEL*

Future use of the CO health risk model can be enhanced by using "fitted" instead of actual distributions, incorporating a "relative exposure factor," and using "at work" population figures in place of resident population figures. Use of fitted distributions might portray annual risk more accurately, especially in cities where the data are not complete. Fitted distributions are now being calculated using UPGRADE. In general, the software divides the data into class intervals (the range and

Table II.
Per Patient CO Related Angina Attack Rates for 9 SMSAs

| SMSA name | average number of attacks/SMSA | average number of attacks at worst location | average number of attacks at worst location/worst day |
|---|---|---|---|
| Philadelphia | 0.0004 | 0.0012 | 0.042 |
| Washington D.C. | 0.0006 | 0.0031 | 0.092 |
| Boston | 0.0008 | 0.0030 | 0.059 |
| St. Louis | 0.0005 | 0.0038 | 0.260 |
| Seattle | 0.0002 | 0.0034 | 0.092 |
| Denver | 0.0037 | 0.0011 | 0.159 |
| Portland | 0.0015 | 0.0048 | 0.025 |
| Phoenix | 0.0013 | 0.0065 | 0.126 |
| Louisville | 0.0024 | 0.0170 | 0.159 |

Table III
Annual Angina Patient Risk Data for 9 SMSAs (thousands)

| SMSA | Population | Total Attacks | CO Related Attacks | CO Related Attacks as % of Total |
|---|---|---|---|---|
| Philadelphia | 4,816 | 111,000 | 16.0 | 0.014 |
| Washington, D.C. | 2,860 | 65,800 | 13.7 | 0.021 |
| Boston | 2,753 | 63,300 | 15.7 | 0.025 |
| St. Louis | 2,362 | 54,300 | 8.3 | 0.015 |
| Seattle | 1,421 | 32.700 | 2.2 | 0.007 |
| Denver | 1,227 | 28,200 | 34.3 | 0.122 |
| Portland | 1,006 | 23,100 | 11.9 | 0.052 |
| Phoenix | 936 | 21,500 | 9.2 | 0.043 |
| Louisville | 826 | 19,000 | 15.1 | 0.079 |

size of the intervals are decided by the analyst), computes the cumulative percent of each interval, plots the percentiles on probability-log scales, and calculates either first, second, or third order polynomial fit through the data. The best fit is determined visually by the analyst. New cumulative percentiles are then derived from the fitted line. These percentiles are translated back into days by multiplying each interval by 3.65. This algorithm "fills in" the missing data when monitoring for the year is incomplete. A comparison of four SMSAs using expected and actual distributions is shown in Table IV.

A relative exposure factor (REF) has been estimated to correct for indoor/outdoor and near source/far source effects on risk rates. Several studies indicate that the actual exposure (weighted for population) to CO is probably closer to 65 percent of the measured levels. This figure was obtained by taking the average of five calculated REFs (Ott, 0.55; Cortese, 0.65; General Electric, 0.50; Travelers, 0.63; and GEOMET, 0.64) and using the following formula:

| % pop in storefront environments | x | Unity | + | Remaining pop | x | the average REF | = | Weighted REF |
|---|---|---|---|---|---|---|---|---|
| (0.15) | x | (1) | + | (0.85) | x | (0.60) | = | 0.65 |

It is assumed that 15 percent of the urban population resides in storefront environments or equivalent, which are exposed to CO levels recorded by monitoring stations.[10] The remaining population is exposed to levels adjusted by the relative exposure factor to yield the weighted relative exposure factor of 0.65.

Taking mobility into account would increase the population at risk as a result of the influx of people into the city during high CO levels. An analysis of Los Angeles yielded an estimate that about 138 percent more people are in the downtown area during working hours than is indicated by the resident census data. Other transportation studies could be evaluated to obtain a range of possible mobility conversion factors. It would appear that the relative exposure and mobility coefficients will offset each other in terms of estimating downtown exposure risks. This belief led the authors to assume a unity coefficient in this study of CO risk.

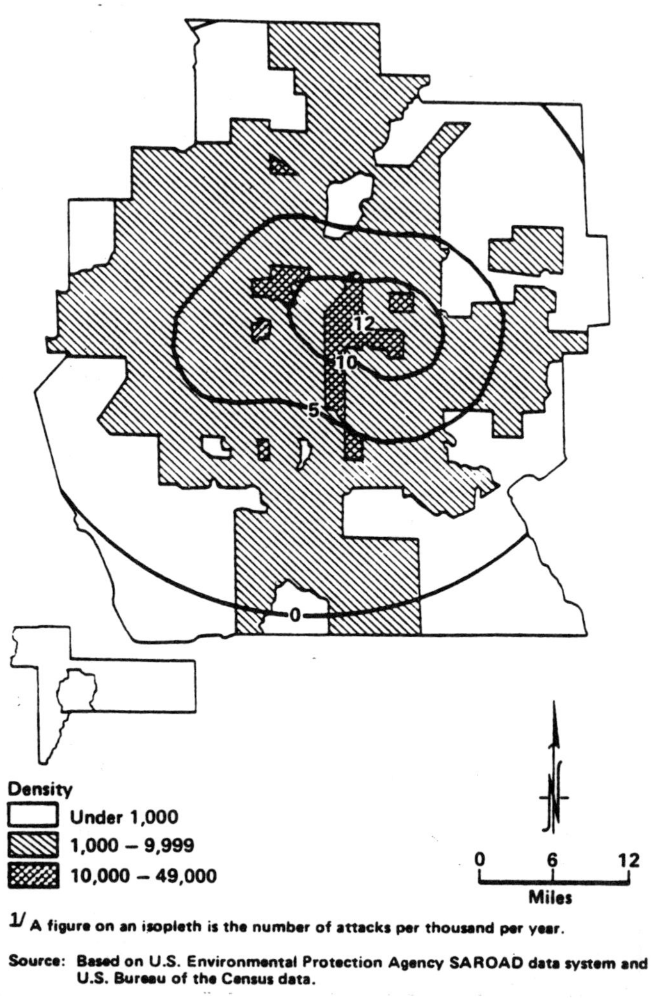


*Figure 4. Projected Risks of CO-Related Angina Attacks, Denver 1985*

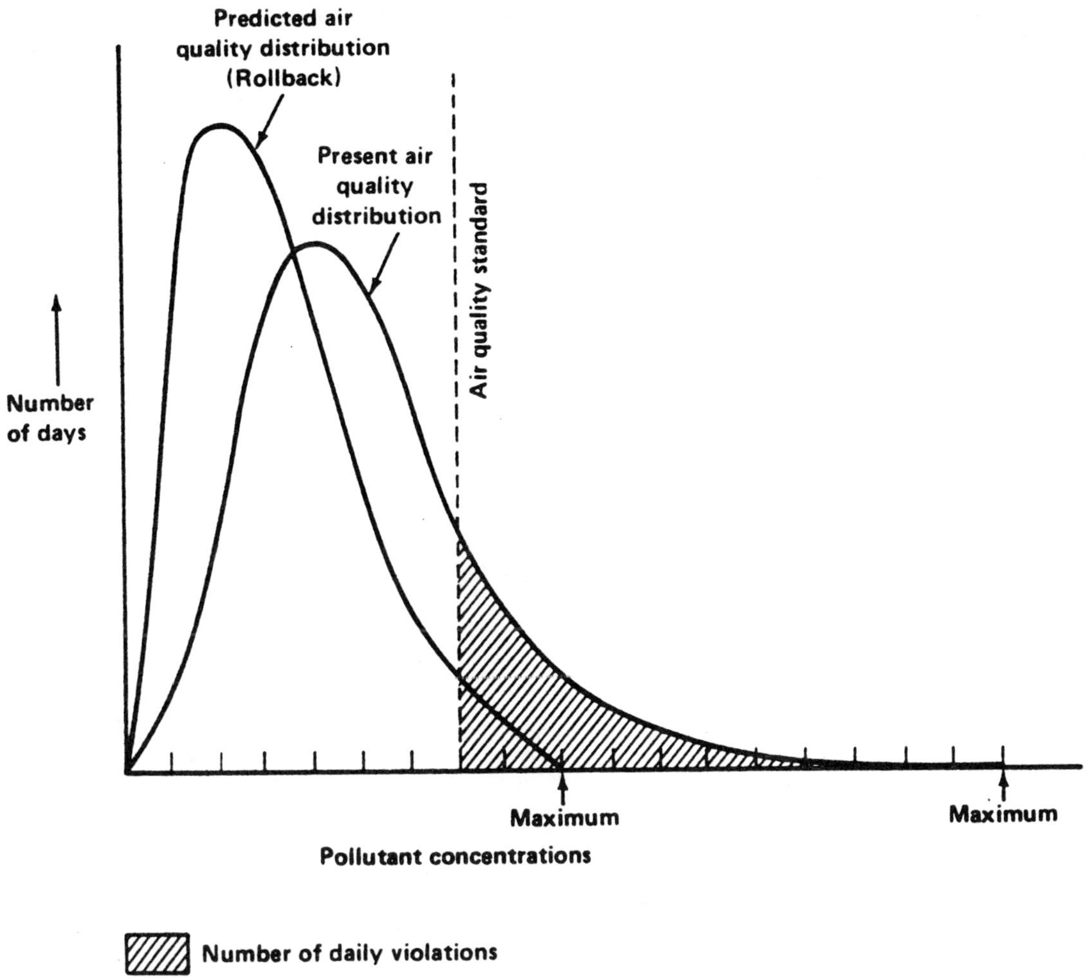


*Figure 5. Observed and Predicted Air Quality Before and After Controls are Applied*

## CONCLUSIONS

The CEQ air pollution health risk model has provided greater insight into actual air pollution health impacts by producing visual and quantitative estimates of health risk and damage. The model is currently being used to analyze those pollutants for which SMSA specific data appear to be adequate vis-a-vis the health-related criteria pollutants. Although the current monitoring coverage for these pollutants is controversial, we find that most of the more polluted cities meet monitoring guidelines.

The damage function describing health risk experienced by stable angina pectoris patients exposed to CO which is used in the CEQ air pollution-health risk model was developed by Jones and Knelson. The results of analyzing nine SMSAs show the most severe stress experienced by stable angina pectoris patients, which is on the maximum exposure day at the site with the highest CO levels, would be approximately 0.26

Table IV
Calculated CO Related Angina Attacks
Using Expected vs. Actual Distributions (thousands)

| SMSA | Total Actual Attacks | Total Expected Attacks |
|---|---|---|
| Denver | 34.3 | 32.2 |
| Philadelphia | 16.0 | 14.4 |
| Portland | 11.9 | 12.0 |
| Phoenix | 9.2 | 12.4 |

attacks in addition to their daily normal attack rates. The average daily risk rate is approximately 0.017 attacks at the maximum exposure location of the nine SMSAs analyzed. Although these figures appear quite low, the authors are not rendering a judgment as to the medical significance of the results. Rather, the authors are providing a methodology for translating clinical findings into estimates of the relative contributions of air pollution to all causes of a particular disease. The medical experts can then, with appropriate recognition of the uncertainties involved, render opinons as to the significance of the air pollution insult.

Besides making this specific risk analysis, the model can make a general exposure analysis simply by calculating person-day exposures to a particular pollutant. There is a maximum total number of exposure days for an SMSA (above a determined pollution level) of 365 multiplied by the total population of the SMSA. This analysis is useful for determining total exposure and making case study comparisons of residual pollution problems associated with various pollution control scenarios. It is hoped that such analysis will replace emissions residuals, percent deviation from a single pollutant concentration, and other vague indicators of public health impacts which appear in most current air quality planning documents.

Further applications of the model can be made by substituting emissions data for ambient air data. This flexibility is useful for making risk analyses when no ambient data exist. When data on emissions and ambient air quality are both available, the mapping package in the risk model will give insight into the spatial relationships of the two.

## *REFERENCES*

[1]Jones, K.H.; R.L. Ferrari. "Development of a Relative Exposure Factor for the CEQ Health Risk Model." Council on Environmental Quality, 1978.

[2]Knelson, J.H. "General Population Morbidity Estimates from Exacerbation of Angina Pectoris Related to Low-Level Monoxide Exposure." Environmental Protection Agency, Health Effects Research Laboratory, Aug. 1975.

[3]Jones, K.H.; J.H. Knelson. "A Continuous Stress Function Relating Ambient Carbon Monoxide Exposure and Angina Pectoris Attacks". Submitted for publication in *Am. J. Public Health.*

[4]Aronow, W.S.; C.N. Harris; M. Isbell; S.N. Rokaw; and B. Imparto. *An. Intern. Med.,* 77;669-676, 1972.

[5]Aronow, W. S., and M. M. Isbell. *Ann. Intern. Med.,* 79:392-395, 1973.

[6]Anderson, E.; R.J. Andelman; J. Straugh; N. Fortvin; and J.H. Knelson. *An. Intern. Med.,* 79-46-50, 1973.

[7]North Atlantic Treaty Organization, Committee on the Challenges of Modern Society. *Air Quality Criteria for Carbon Monoxide.* Government Printing Office, 1972, p. 7-13.

[8]Haak, E., Health Effects Research Laboratory, Office of Research and Development, Environmental Protection Agency, 1978.

[9]Jones, K.H., *CEQ Projections of Ozone Air Quality in 1990.* National Technical Information Service, Springfield, VA, 1977.

[10]Cortese, A.D. Dissertation, Harvard School of Public Health, Cambridge, MA, 1976.

# *Spatial Data Integration Requirements For Earth Resources Applications*

*by Ellen M. Knapp and Martha R. Szczur*

## *SPATIAL DATA INTEGRATION — OVERVIEW*

### Background

Increasingly, data producers are providing their spatial and geo-referenced data in both analog and digital format. In addition, data users in government and private industry are converting existing data to digital files and developing geo-referenced data-bases. Geo-referenced data include spatial data from existing graphic or image sources (both planar and three-dimensional), nominal, descriptive, and attribute data, tabular data, and textural information.

Initially, the major requirements for the development of digital data-bases were the software and hardware capabilities necessary to support data encoding, editing, storage, retrieval, and display. The application of data management software became necessary to provide for efficient data storage, retrieval, and update as well as data independence, security, and integrity. More recently, focus has been centered upon the need for spatial data manipulation and analysis capabilities.

Data manipulation and analysis include a wide range of software and hardware capabilities designed to facilitate the processing of a single data type, multiple data types within a single data-base, and multiple data-bases. Data manipulation operations include reclassification and aggregation of attribute data; rotation, translation, and scaling of coordinates; conversion of geographic coordinates to specified map projections; distortion elimination; centroid and line allocation; and data structure conversions. Data analysis procedures include measurement, statistical analysis, and spatial analysis. Several comprehensive publications are available which provide detailed descriptions of these techniques as well as existing software and hardware capabilities.[1 2 3]

For most operational data users, currently available data processing and spatial analysis technique are being implemented. Concurrent with the existing technology utilization effort is the development of improved capabilities as new data sources become available, digital data-bases more extensive, analysis requirements more sophisticated, and timely information for both researchers and decision-makers in greater demand. Research and development programs designed to address new and improved techniques exist within the federal government (e.g., NASA; U.S. Geological Survey; and the Oak Ridge, Berkeley, and Los Alamos National Laboratories), universities (e.g., Harvard University, State University of New York at Buffalo, University of California, University of South Carolina, and George Washington University), and private industry (primarily corporations providing software and hardware to support data producer and users). It is significant to note, however, that many of these programs are data-specific, project-specific, or hardware-specific; furthermore, adequate publication, documentation, transferability, and software development standards are conspicuously lacking.

### Definitions

This paper is concerned with a set of specific data manipulation and analysis techniques referred to as spatial data integration. Spatial data integration is defined here as the process of combining multiple spatial and geo-referenced data types and providing for their subsequent storage, retrieval, analysis, and display. The need for spatial data integration was well established, before the implementation of automated procedures, by data users who required multivariate statistical or spatial analysis. For analog data users, this required the acquisition of multiple data types from variety of data sources (existing maps and charts, aerial photographs, and geo-referenced tabular files); preprocessing (data interpretation, reclassification, remapping, and registration); and data integration (map overlay and attribute data correlation). Some of the many problems associated with this process are data identification and acquisition; data comparability and compatibility (temporally and spatially); data accuracy and reliability; data storage and management; and integrated and interpretive data analysis. (See *Geographic Information Systems, Methods, and Equipment for Land Use Planning,*[1] for a review of manual data processing techniques.) This paper is primarily concerned, however, with automated spatial data integration.

Automated methods were originally developed to duplicate manual procedures and have more recently been expanded to provide data integration capabilities which are not feasible, or not possible, using manual techniques. For data users with large volumes of multiple data types, automation can make a significant contribution to data processing in terms of the cost and time required for data management, data retrieval, data analysis, and data display in addition to the increased capabilities provided. Unfortunately, however, many of the problems associated with manual data processing in conjunction with complex problems specific to digital data, software, and hardware, make automated spatial data integration less than a straightforward set of procedures. The final section of the paper describes some of these problems and the steps being taken to address these issues by NASA, other agencies, academia, and private industry.

Automated spatial data integration (SDI) has been defined as a set of procedures which are related to software/hardware capabilities and data characteristics. Software and hardware capabilities include data manipulation algorithms, application packages, man-machine interface, data management sytems, CPU and I/O requirements, storage type and limitations, and available peripheral devices. To provide for efficient integration, software should be data-independent, with software package and data systems linked to provide for data interchange, and (if possible) project-specific systems linked to general purpose processing systems. General characteristics of spatial and geo-referenced data include the medium on which the data are recorded; the structure and format of the data; temporal characteristics (date, currency, and uniformity); resolution and scale; volume; projection and/or coordinate reference scheme; quality, accuracy, and precision; obervational criteria (direct measurement, categorized, or interpreted); measurement scale (nominal, ordinal, interval, or ratio). Software and hardware capabilities will be hereafter discussed as automated data processing (ADP) issues; those integration issues which are associated with the data characteristics will be referred to as discipline issues. Admittedly, ADP and discipline issues are interrelated. There are, however, important distinctions. Data-independent software and hardware features allow for the integration of spatial and geo-referenced data in both appropriate and inappropriate ways. The following is a simple example: using the same techiques, it is possible to display total population counts by census tract, or population density by census tract. Many users are not aware of the major differences in the interpretability of these two output products. The techique which allows for the retrieval and display of the data is an ADP capability; which measurement scale is appropriate for the display is a discipline issue. Unquestionably, the discipline issues are less well defined. Many are well known to experts and unknown to most data users; others have been identified by experts and are little understood; and undoubtedly some are as yet unidentified. However, the need for and the use of SDI techniques will not "wait for" a full set of ADP capabilities and completely defined discipline issues. Data users will continue to apply existing methods, develop and implement new techniques, and should be as informed as possible of the discipline issues related to the analysis performed.

In order to provide more efficient spatial data integration, a project was initiated at GSFC to analyze integration requirements and current capabilities. The following section describes the objectives and methodology of the project.

## *SDI PROJECT — OBJECTIVES AND METHODOLOGY*

NASA is a major producer of spatial data, primarily in digital format, for NASA researchers and non-NASA data users. These data include a broad range of measurements related to atmospheric effects, weather and climate, hydrology, limnology, oceanography, geophysics and geodynamics, and land resources. Researchers at GSFC analyze these data for studies in one of six space applications areas: 1) earth resources detection and monitoring, 2) earth dynamics monitoring and forecasting, 3) ocean condition monitoring and forecasting, 4) environmental quality monitoring, 5) weather observation and forecasting, and 6) climate research. Many of the ongoing research projects are conducted with a single remotely densed satellite data type; however, most require ancillary (additional, related) remotely sensed or ground-based data. As studies have progressed from sensor design, data collection, mapping, and monitoring to predictive modeling and forecasting, the need for ancillary data and automated spatial data integration capabilities has rapidly increased.

Broadly stated, GSFC has a requirement to integrate remotely sensed data with ancillary data, share source data and derived information, and to provide for storage, retrieval, analysis, and display of these data to meet the researcher's information needs. A study was initiated at GSFC to analyze the SDI requirements within the space applications areas and evaluate current capabilities for spatial information processing. The objectives of this study are as follows:

- Define a broad set of requirements and standards for interfacing remotely sensed satellite data and ancillary data types.
- Develop and evaluate approaches to integrating spatial data and interfacing data handling resources.
- Identify and prioritize project-independent SDI research needs (both ADP and discipline issues).
- Recommend measures to expand users' analytic tools, enhance data integration capabilities, and provide efficient data interchange.
- Promote information dissemination.

Recognizing that certain ADP and discipline oriented requirements are project or data specific, the study has been divided into two tasks. The first task is to perform the requirements analysis and capabilities evaluation for a broad range of application research projects with the objectives focused upon project-independent techniques, systems, and discipline issues. The second task is to examine, in detail, the ADP requirements and data characteristics of two projects within the earth survey applications area, focusing upon specific data types, model developments, and research needs. The detailed task is designed to provide valuable information to the study team for developing general design approaches to enhance SDI capabilities and for identifying and prioritizing unresolved research topic areas.

The methodology developed for this study is applicable to both the broad-based and the in-depth tasks. Procedurally, there are three major steps, project identification, data definition and requirements analysis, and evaluation of current capabilities.

### Project Identification

The first step was the identification of research projects within the space application areas with existing or proposed SDI requirements. It is significant to note that SDI requirements do not necessarily imply the use of multiple data types; for example, a researcher performing temporal analysis of a single data type, where the data are not temporally registered (i.e., LANDSAT data), requires SDI techniques. Therefore, the user's information requirements needed to be evaluated before the applicability of a project to this study could be determined. The volume of information necessary to meet the objectives of both the broad and the in-depth tasks prohibited the incorporation of all GSFC applications research projects. A representative set from four of the space application areas was selected. Table I lists the four application areas and examples of research projects within these categories. The criteria for the development of a representative set were chosen to include a wide range of remotely sensed satellite data and ancillary data types; current and proposed projects; projects with well established SDI methods as well as those with ill-defined integration requirements; and projects which are currently utilizing one of several software packages and spatial data processing systems available on computers at GSFC (both large-scale and mini-computers).

In order to obtain information related to each project, on-site staff (GSFC or contractor) were identified. Frequently, project staff were so specialized that two or more members of each team had to be contacted to gather the information necessary for the data definitions, SDI requirements analysis, and evaluation of current capabilities. Project identification tables were prepared that provide a task generic name and task identification number to which all subsequent tables are referenced.

### Data Definition and Requirements Analysis

Having identified the application research projects, appropriate staff members were contacted to obtain information related to spatial and geo-referenced data types and SDI requirements. Detailed characteristics of each data type were necessary to assess the feasibility of SDI requirements and the applicability of existing capabilities. Data definition tables were prepared to provide a categorization of the following characteristics by project and data type:

- Data source: the data producer for source data; the source data type for derived (interpreted or processed) data.
- Spatial data type: digital — medium on which the data are recorded, format, structure [point, line, area (grid or other)], surface; analog — map, chart, photoproduct, controlled or uncontrolled, structure.
- Geo-referenced data type: geo-referenced scheme (i.e., the structure of the spatial data type to which it is referenced).
- Date(s); resolution and scale; sampling frequency; volume (record volume, data volume, density); projection and coordinate reference scheme; accuracy or quality code.
- Observational criteria: direct measurement, grouped or classified, interpretive.
- Project spatial coverage.
- Multi-sample (temporal) data: interval, number of samples, temporally registered or unregistered, registration accuracy.
- Multispectral data: number of samples (channels), spectral coverage by channel, radiometric resolution (gray level range), accuracy of registration.
- Measurement scale: nominal, ordinal, interval, ratio.

The data definition tables are later cross-referenced with the software and hardware characteristics tables to determine (by data type and project) which systems are currently used to process specific data types (e.g., LANDSAT data-Interactive Digital Image Manipulation System — HP3000 Series II). The objectives and potential results of cross-referencing the tables will be discussed in greater detail in the following section.

TABLE I
Space Applications Areas

I. EARTH RESOURCES DETECTION AND MONITORING
- Land Cover/Land Use
- Agriculture
- Forestry
- Geology
- Hydrology, Limnology

II. EARTH DYNAMICS MONITORING AND FORECASTING
- Geodynamics
- Geophysics

III. WEATHER OBSERVATION AND FORECASTING
- Severe Storms and Local Weather Research
- Global Weather Research

IV. CLIMATE RESEARCH
- Global Biomass Monitoring
- Atmospheric Constituents
- Energy Budget

One of the stated objectives is to define the scope of SDI requirements within the applications areas. After identification of research objectives and data characteristics, the status of SDI requirements for each project is determined (i.e., integration techniques currently utilized, proposed, or under consideration). For projects currently utilizing SDI procedures, the following information is obtained: software and hardware supporting data analysis, data types integrated, products resulting from the integration (e.g., statistical summaries, models, and graphics), additional SDI capabilities required, and problem areas. Where the implementation of integration procedures is proposed or under consideration, the requirements are addressed from both the ADP and discipline perspectives. The ADP requirements are then categorized as project-dependent or independent and the discipline issues as data-dependent or independent. For the broad level requirements analysis, the project- and data-independent needs are of primary concern. This information will be used to develop general design approaches which will benefit a wide range of satellite data users. Conversely, the two in-depth analyses focus on project- and data-dependent requirements.

### Current Capabilities

A prerequisite to the development and evaluation of data integration approaches is the identification of available data manipulation and application software which is currently using the defined data as a source. GSFC has an abundance of sophisticated information extraction software packages which are used as analytical tools in their specific research and development work. In Table II these current data analysis capabilities are grouped into five general categories, and examples of software packages available for earth resources applications are given for each group.

Despite such an assortment of existing spatial data extraction packages, most of the systems at GSFC were purchased or designed for a specific project- or data-dependent service. Consequently the ability to convert between and to combine different spatial data types is a significant problem. The systems are independent of one another and implemented on a variety of different computers.

This complicates data conversion and transferability by adding a new set of limiting elements: compatibility of operating systems, input/output devices, data storage structures, large-scale versus mini-computer capabilities and other machine-oriented system elements. Within each of the existing systems, the ability to input and integrate different spatial data conversion capabilities is inadequate. Even the commercial data base management systems are oriented towards homogeneous data structures and do not have efficient mechanisms for handling spatial data.

Although project independence certainly has its merits, it tends to limit information exchange between data users. Because of the current requirement for spatial data integration techniques within many of the applications research projects, a complete identification of the available data manipulation and analysis software becomes necessary. In addition, this evaluation will allow an accurate survey to be made of how the current capabilities specifically do or do not meet the defined SDI requirements.

In order to evaluate the current capabilities, a set of basic package characteristics was identified. For each software system which supports the processing of previously defined data types (refer to Data Definition and Requirements Analysis section), the characteristics were gathered by documentation review and interviews with appropriate GSFC and contractor staff. The following categories were included in the software package survey:

- Name of package.
- General description of software's function.

TABLE II
Spatial Data Information Extraction Capabilities

| CATEGORY | EXAMPLES |
|---|---|
| Image Processing Systems | JPL's Video Image Communication and Retrieval System<br>ESL's Interactive Digital Image Manipulation System<br>Atmospheric and Oceanographic Information Processing System<br>GE's Image 100 |
| Data-Base Management Systems | HP's QUERY<br>Climate Data Base Management System |
| Application-Dependent Modeling Programs | Hydrology Model<br>Climate Model |
| Standard Statistical Packages | SAS — Statistical Analysis System<br>BIOMED<br>SPSS — Statistical Package For the Social Sciences<br>MINITAB |
| Geocoded Information Processing Systems | JPL's Image Based Information System<br>ESRI's Geographic Information System<br>Domestic Information Display System |

TABLE III
Interface Classification

| | |
|---|---|
| Link Not Established | Link not established and not required<br>Link not established and required and feasible<br>Link not established and required and not feasible |
| Direct Link Exists | Dual Port Disk<br>High Speed I/O Lines ( >9600 bits per second)<br>Low Speed I/O Lines (i.e., telephone lines via modem) |
| Indirect Link | Required stand-alone preprocessing/conversion step<br>Output product becomes direct input to other package |

- Operational information: system manager, computer facility, average monthly usage, documentation titles, current user types.
- Data information: type of input data (based upon categorization in the data definition phase), type of output produced, brief description of data file storage used within package, storage size requirements.
- Software characteristics: level of man-machine interface(i.e., menu-driven, batch, etc.), software size, structure, languages, type of I/O handler, transferability rating.
- Hardware characteristics: definition of processing and input/output devices utilized by the software package (i.e., display terminals, plotters, array processor, etc.).
- Interface characteristics: definition of available computer-to-computer links used to transfer data between software packages, definition of available conversion software enabling data exchange with other software packages (if available, definition of the original and converted data types and the conversion technique.).

### Integration Evaluation

Once the data requirements anlaysis and current capabilities identification are completed, it becomes feasible to develop and evaluate approaches to integrating spatial data and interfacing data handling resources. To aid in this integration evaluation, a program-to-program link matrix and data usage tables were developed.

The program-to-program link matrix is a table which identifies the interface relationship between each of the defined existing software packages. The linkage interfaces are classified by a two-level hierarchical structure and Table III defines the preliminary classification categories.

As an additional tool in defining the ADP issues, the data usage table supplies a cross-referencing of defined data types versus defined software packages. For each matrix element, a list of the applications research projects that use (or would like to use) this combination of data and software is defined. By evaluating this table, one can determine the combination of data types which are currently processed or being considered for processing by each of the GSFC projects. Besides serving as a classification table of existing data usage, this table pinpoints the versatility and limitations of existing packages and provides input to the design approaches for integration enhancements. By referencing the program-to-program link matrix and the data usage table together, the data integration requirements are categorized into two groups: 1) the ADP integration areas of interest (i.e., the "links not established and required and feasible") which will need further evaluation to determine if the data handling resources should be interfaced; and 2) the "links not established and required and not feasible," which will need further examination before being identified and prioritized as a research topic area.

ADP integration techniques will vary in scope, and preceding the development of an implementation design, each job must be evaluated and its priority established. A set of priorities will be developed based upon the following factors: degree of need, difficulty level, implementation cost, and time required for implementation. Essentially the same set of factors will be applied to the research topic areas with obvious omission of factors relating to implementation.

Once the integration processes have been "sorted," the development of general design approaches to implement the interface functions can be undertaken. The range of the data integration processes will include the following design categories:

- Inclusion of new spatial data as a direct input into an existing software package. This would involve the addition of new functions to process the data and integrate them with other data.
- Stand-alone conversion process developed to convert/manipulate data into an accepted format for input to existing software packages.
- Interface/integration software designed to develop data exchange between existing software packages.
- New processing capabilities designed to handle spatial data integration requirements.

When applicable, several different design approaches will be developed, which will propose different levels of implementation effort. A direct data I/O interface line between two software systems might be proposed as the most efficient method of linkage. An alternative or preliminary method might involve hand carrying a data tape across the GSFC facility and submitting it as input to a conversion routine, which will then produce a data tape that can be used as input to another software system. Depending on the user's information requirements, either design could be the optimum approach.

It is significant to note that one additional input will be used when developing the general integration designs. Several other government agencies which are known to have spatial data integration requirements (e.g., DMA, DARPA, USGS) will be contacted to determine how they are approaching system/data interfaces in-house. Whether these approaches would be applicable to GSFC applications and environment is another factor to be considered during the integration evaluation.

The following section provides a brief description of SDI requirements and capabilities within the context of an ongoing earth resources application research project.

## *THE INTEGRATION OF LANDSAT AND ANCILLARY DATA FOR IMPROVED URBANIZED AREA DELINEATION*

As part of the decennial census conducted by the U.S. Bureau of the Census, urbanized area (UA) boundaries are revised to reflect changes in population and land use that occur with urban expansion. Urbanized area boundaries are used as a funding criterion for a number of federal programs, including General Revenue Sharing, Urban Mass Transit, and the Federal Highway Act.

The following procedures are used by Bureau of the Census geographers in urbanized area delineation:[4]

- Gathering information about new residential structures outside of the existing urbanized area boundary.
- Designating a maximum possible extent for the urbanized area referred to as the fringe zone.
- Dividing the fringe zone into enumeration cells.
- Obtaining population counts and area measurements of the enumeration cells.
- Computing the density of each enumeration cell and then classifying the cell as urbanized or not based on the population requirement of 1000 persons per square mile.

These procedures are both costly and time consuming; furthermore, they are often limited in accuracy by the timeliness of information about new residential structures. In addition, new legislation and the rate of expansion of urban areas may require more frequent boundary updates.

The search for improved UA delineation techniques led to a joint project between the U.S. Bureau of the Census and NASA designed to assess the utility of LANDSAT digital data for faster, more cost effective delineation of urbanized areas. This project, the Census Urbanized Area Applications Pilot Test (APT) Project, is now underway and a variety of LANDSAT analysis techniques are being developed and evaluated.

This new approach to UA delineation treats expansion as a function of urban land cover. Land cover information from LANDSAT data can be used to provide a timely, objective, repetitive, uniform data source to supplement conventional data sources. The goal of the Census APT is to develop and evaluate LANDSAT data analysis techniques which will contribute to effective delineation of urbanized area boundaries. A variety of analysis techniques and output products is being developed and tested for urbanized areas in selected standard metropolitan statistical areas (SMSAs). Several approaches and combinations of approaches are being evaluated, including the following:[2] photointerpretation of raw LANDSAT data; photointerpretation of digitally enhanced LANDSAT data; digital classification of LANDSAT data for urban land cover; multitemporal urban land cover change detection; LANDSAT inputs to a geographic information system.

Detailed descriptions of the first four approaches can be found in previous publications.[4,5,6,7] The purpose of this section is to summarize current efforts addressing the fifth approach — LANDSAT inputs to an automated geographic information system (AGIS).

### General Concepts

There are three primary approaches to interfacing LANDSAT and ancillary data: manual overlay for enhancement of output products, digital overlay for statistical summaries by unit area, and interface with AGIS. Manual and digital overlay techniques have been developed and tested and are being evaluated by Bureau of the Census personnel. Technique development for integrating LANDSAT and AGIS technologies is currently being investigated. An automated geographic information system is an information system which has as its primary source of input a base composed of data referenced by geographic coordinates and in which a major part of the processing is done with a digital computer[5]. An AGIS should include software for encoding and input processing of spatial and nonspatial data; data management; manipulative and analytic operations; data retrieval; and display of maps, graphs, and tabular information on a variety of output media (Figure 1).

### Objective and Methodology

The primary objective is to improve the LANDSAT land cover classification procedures and products. To examine the utility of AGIS processing techniques, this objective has been redefined as a series of specific research tasks:

1. Develop remotely sensed satellite data processing algorithms which incorprate ancillary data.
2. Assess the optimum resolution of digital terrain data for surface cover mapping improvement.
3. Evaluate LANDSAT data inputs to urban growth models.
4. Develop improved accuracy assessment techniques.
5. Enhance data display capabilities.

A project was designed to address these objectives and a test site in the urban fringe southeast of Denver, Colorado, was selected. The remotely sensed satellite data and ancillary spatial and nonspatial data acquired for this project are listed in Table IV. These data were acquired from NASA, the Bureau of the Census, USGS, Soil Conservation Service, and state and local planning agencies. They represent a wide range of data characteristics — analog and digital, spatial and nonspatial, scale and resolution, geometrically corrected and uncorrected, data structure and format. During data acquisition, relevant information for each data type was obtained which describes the accuracy, precision, currency, and other characteristics of the data. Each spatial data type is referred to as a spatial data layer. Within each spatial data layer there are four possible types of geographic entities which must be encoded: points, lines, surfaces, and area-enclosing lines or polygons. From each spatial data layer, then, the geographic entities are encoded (or input directly in digital format), using one of two basic techniques. The entities may be generalized and stored

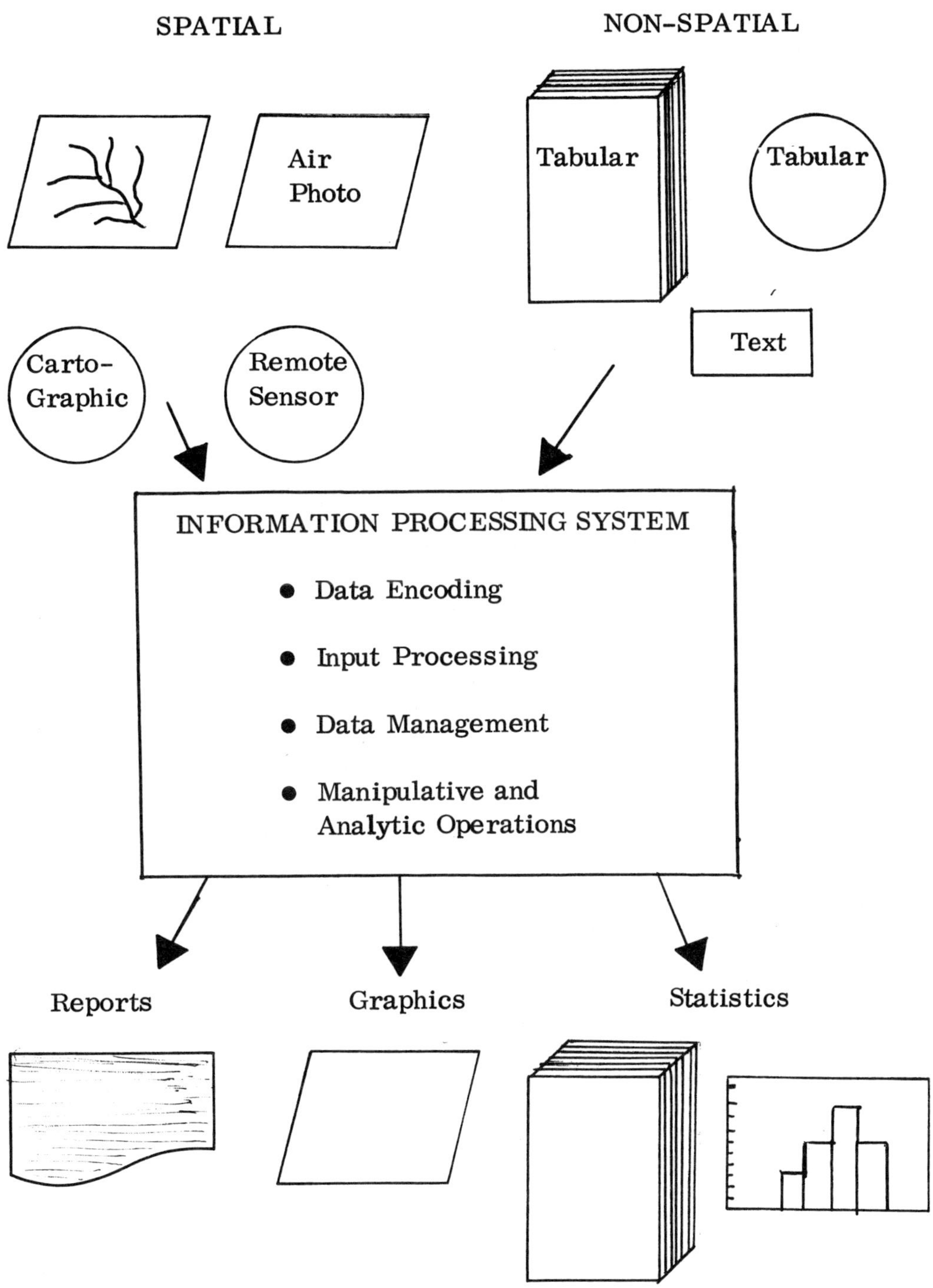


*Figure 1. AGIS Functions*

by the use of grid cells as a location identifier; however, this generalization incurs the loss of geographic specificity. To retain more precise locational information, entities may be encoded and stored using an x,y coordinate description. Grid cells may be regular area cells, irregular (hierarchical) cells, irregular triangulated cells, or some other grid representation. The x,y coordinate data are encoded and, after input processing, stored as points, lines, networks, surfaces, or areas (polygons). The advantages and disadvantages of grid versus x,y coordinate structures have been well documented. The basic issues are data volume (or storage efficiency), data manipulation (or processing efficiency), data accuracy, and data display. For maximum flexibility, a system should be able to accept, store, manage, retrieve, and display both types of data and provide for conversion from one data structure to the other. Thus, data could be stored in their "natural" form to optimize geographic specificity; data could be converted to grid format for analysis and manipulation; and data stored in

TABLE IV
Spatial and Non-Spatial Data Types

**Remotely Sensed**
LANDSAT MSS and RBV
SEASAT SAR
LANDSAT-D Simulator (Aircraft)
High Altitude Color-IR (NASA)
Low Altitude B/W (USDA)

**Census**
Urban Atlas Files
Urbanized Area Boundary
Metropolitan Map Series
Census Summary Data

**USGS**
Topographic Maps
Digital Elevation Tapes (TOPOCOM and DEM)
Thematic Maps

**Soil Conservation Service**
Survey Reports
Soil Maps

**State and Local Planning Agencies**
Land Use
Zoning
Ownership

x,y coordinate format could be converted to grid format at a user-specified cell size.

In order to provide AGIS capabilities, GSFC acquired a commercially available system from the Environmental Systems Research Institute (ESRI) and has implemented the software on an IBM 360 and HP-3000 Series II. The ESRI-AGIS provides all of the capabilities listed above, including the ability to handle both x,y coordinate and grid structured data. It does not, however, include image-processing functions. Remotely sensed satellite data would have to be preprocessed using one of the available image analysis systems and then converted to ESRI grid format. As stated previously, one of the objectives of this project is to develop image classification algorithms which incorporate ancillary data. In order to meet this objective, the AGIS and image analysis systems would have to be linked to provide not only format conversion operations but also the capability of operating on the various data types with applications programs within each system.

As part of the Census APT, researchers have been using two image-processing systems at GSFC: the video image communication and retrieval (VICAR) system on the IBM 360 (developed by the Jet Propulsion Laboratory) and the interactive digital image manipulation system (IDIMS) on the HP-3000 (a turnkey system developed by ESL). Links between the ESRI-AGIS software and the two image analysis systems have been designed. The following section describes the link between the ESRI-AGIS and IDIMS.

## *IMAGE ANALYSIS AND GIS TECHNIQUES*

An image-processing system is developed for manipulation, analysis, interpretation, and display of digital image data. In the IDIMS system these functions include the following:

- Image enhancement — smoothing, edge detection, geometric correction, reduction, expansion, and filtering.
- Analysis — classification, texturing and arithmetic computation.
- Display control capabilities for visual interpretation manipulation.
- Statistical reporting — histograms, maximum/minimum values, mean standard deviation.
- Image data registration and overlay.
- Output generation — soft-copy and hard-copy (electrostatic printer/plotter, film recorder).

Image data structure (picture elements or pixels) is ideally suited for these types of manipulation and enables image analysis to be effectively performed (e.g., Boolean combination, local operations, parallel processing). To support IDIMS functions, ESL developed the Geographic Entry System (GES) to allow for encoding of x,y coordinate data, and the Earth Resources Inventory Systems (ERIS) to provide a general purpose statistical package.

A geographic information system, (GIS) is designed for manipulation, analysis, interpretation, and processing of spatial and geo-referenced data. The ESRI software provides these capabilities for both x,y coordinate and grid structured data sets. The polygon information overlay system (PIOS) is for processing of the x,y coordinate data and includes functions for encoding, editing, analysis, modeling, and map generation of two-dimensional data. The GRID system provides similar functions for both two and three-dimensional grid data. The GRIPS package allows for conversion of x,y coordinate to grid data structure. Both the x,y coordinate and grid data can be coded with a single attribute in a single variable file (SVF) or with multiple attributes in a multiple variable file (MFV). See Figure 2.

There are numerous similarities in the functions of image-processing systems and geographic information systems; however, each system was designed and developed for a specific data- or project-dependent purpose. For many applications research projects it has become necessary to combine image and nonimage spatial data and to process these data with applications programs from both types of systems. In order to develop a fully integrated information processing system, a software link between the ESL system and the ESRI system was designed and named LINKPAK (Figure 2). Interfaces currently exist between the following sets of software packages: PIOS to GRID, GES to IDIMS, IDIMS to ERIS. There is no way, however, to exchange data between PIOS/GRID and GES/IDIMS/ERIS. Realizing that the link package must be cost effective and an efficient method of communicating between the two systems, the data exchanges were only defined at the processing points where the two systems' structures were most analogous. This enables all data types from either system to be integrated into the other system's process flow by using one of the LINKPAK conversion functions or by using a combination of the existing system interfaces with a LINKPAK function.

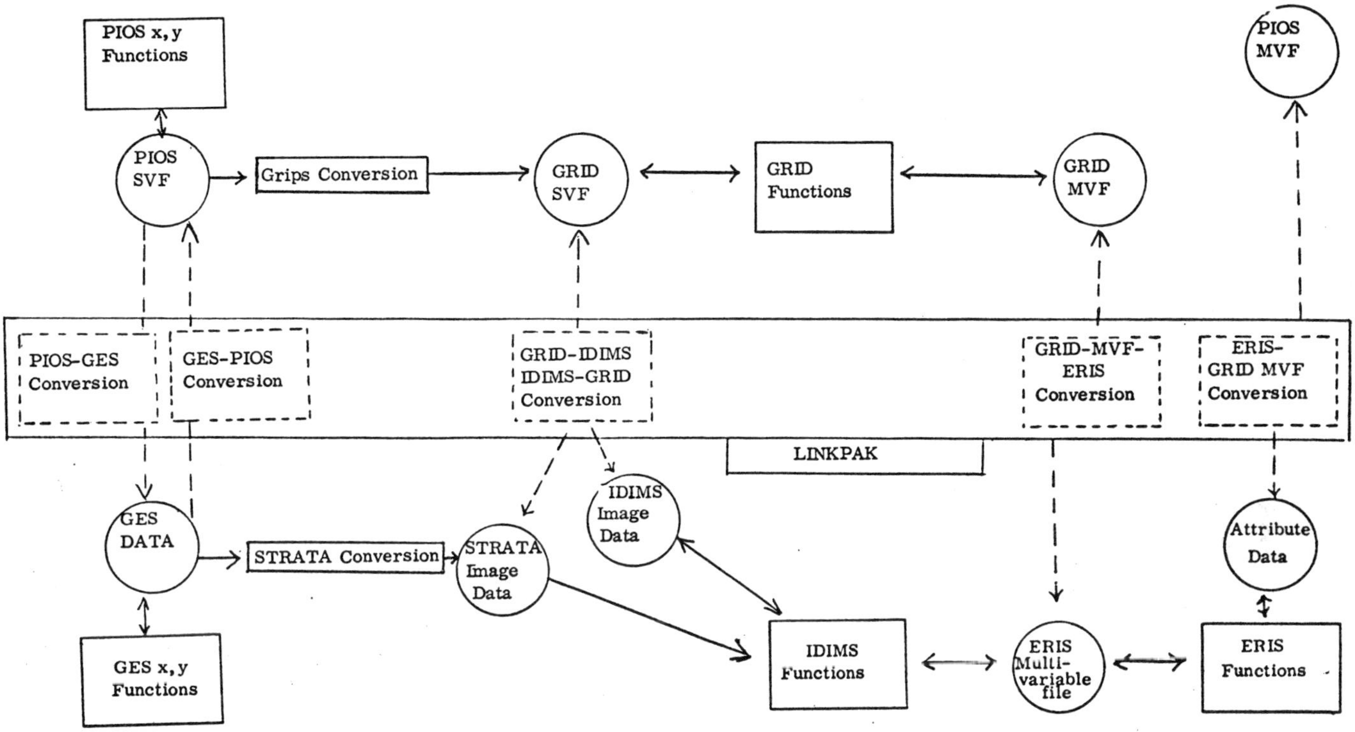


*Figure 2. LINKPAK Functional Overview*

Because of the modular structure of the LINKPAK, the software can be designed and implemented in phases. The data interfaces were assigned priorities based on which links would be immediately utilized and on the complexity of the linkage software implementation. The scope of the LINKPAK can also be implemented in sequential stages. The completed interface system will include an interactive menu-oriented driver which would prompt the user about which conversion is desired and then request the necessary input parameters. This saves the user from having to refer to a separate user's guide just for the LINPAK process. The initial system will be a collection of independent conversion functions, each of which is handled as a separate processing job.

By integrating the image-processing system, geographic information system, and statistical package, a comprehensive information-processing system is created which is able to function with image, spatial (x,y coordinate and grid), and nonspatial data in point, line, area, and surface structures. In addition, the format conversions and data integration capabilities required to combine the different data types will be facilitated. For example, prior to the systems linkage, each new digital data type acquired had to be converted to ESL IDIMS format and separately to ESRI PIOS or GRID format. Following the systems interface, these conversion routines need be programmed only once for each data type. It is significant to note, however, that this systems integration is project-specific and only a small step in comparison to the requirements of GSFC researchers using many other independent software packages on other host computers.

## SUMMARY

The need to integrate spatial and nonspatial data types will continue to increase as GSFC researchers develop new applications of satellite remote sensing. Although in-house data-processing facilities are extensive, a number of data handling and analysis issues will have to be addressed before full use can be made of these automated procedures. The SDI project has been designed to provide the information necessary to prioritize these issues and to develop approaches which will expand users' information extraction capabilities. NASA data users rely, in part, on university research programs in the areas of image-processing, spatial analysis, data base management, and model development. Timely knowledge of academic efforts, as well as information related to other agencies and their approaches, will be required in order to meet these objectives.

## *REFERENCES*

[1]Calkins, H. W. and R. F. Tomlinson. *Geographic Information Systems, Methods, and Equipment for Land Use Planning.* International Geographic Union Commision on Data Sensing and Processing, Ottawa, 1977.

[2]Power, Margaret A. *Computerized Geographic Information Systems: An Assessment of Important Factors in Their Design, Operation, and Success.* Center for Development Technology, Washington University, St. Louis, 1975.

[3]Kennedy, Michael and Charles R. Meyers. *Spatial Information Systems: An Introduction.* Urban Studies Center, Louisville, 1977.

[4]Walthall, C. L. and E. M. Knapp. *The Classification of LANDSAT Data for the Orlando, FL, Urban Fringe Area.* Computer Sciences Corporation, CSC/TM-78/6088, 1978, p. 1-1.

[5]Christenson, Jerrold W.; David L. Dietrich; Henry M. Lachowski; Mark L. Stauffer; and Richard L. McKinney. "Urbanized Area Analysis Using LANDSAT Data" (paper presented at the International Users' Conference on Computer Mapping Software and Data-Bases, Harvard University, 1978, p. 6)

[6]Christenson, Jerrold W.; James B. Davis; Valerie J. Gregg; Henry M. Lachowski; and Richard L. McKinney. *Landsat Urban Area Delineation.* NASA, GSFC, 1977.

[7]Stauffer, M. L. and R. L. McKinney. "LANDSAT Image Differencing as a Change Detection Technique." *Proceedings of the American Society of Photogrammetry,* Little Rock, AR, 1977.

# The Use of LANDSAT, Digital Terrain, and Ground Sample Data as Inductively Derived Information Input to a Multi-Spectral Classifier for Wildlands Mapping and Inventory*

*by Ronald G. McLeod and Thomas L. Logan*


### ABSTRACT

An approach to natural resource inventory and mapping of wildlands (i.e. forests and rangelands) is being developed at the Jet Propulsion Laboratory. The basic assumption is that spectral reflectance patterns are related in a rational manner to resource characteristics and that from this relationship practical classification schemes can be developed. Multispectral LANDSAT imagery is used in conjunction with digital terrain data, and a ground sampling strategy has been designed to include all wildlands cover types as they vary over differing terrain situations. The goal of providing accurate and cost-effective resource managment information has been accomplished by applying inductively derived information to a multispectral classifier.

In the first case study, estimates in the Klamath National Forest of timber volume by species are obtained through unsupervised clustering of LANDSAT spectral data with a synthesized texture channel and Forest Service ground sample data. Slope, aspect and elevation channels derived from registered terrain data are calibrated by ground sample information to predict species proportions through a trend surface model. Volume estimates and predicted species are then merged and aggregated to yield timber volume by species within management areas.

In the second case study, approximately 3,000 miles of two-pixel wide transects were extracted from a ten scene LANDSAT digital mosaic as a representative sample of the desert rangelands of California being inventoried by the Bureau of Land Management. The sample set was input to an unsupervised clustering algorithm which was then used to yield one hundred basic resource classes. Final resource class assignment for soils and vegetation types was aided in a post classification sense with digital terrain information registered to each 1° x 1° segment extracted from the digital mosaic. Additional ground sample information was merged with approximately 500 low altitude air photo transects to derive standing biomass and range carrying capacity values which later are to be aggregated according to land ownership and leased grazing allotments.


### CONCEPTS

The approach to natural resource inventory and mapping of wildlands (i.e., forests and rangelands) integrates LANDSAT multispectral data and ancillary data types. This is based on the wildlands classification concepts developed by Yapp (1922) for vegetation, Milne (1935) for soils, Christian (1958) for land units, and summarized by Mabbutt (1968). All of the researchers have recognized the complexity of the natural environment and the need to inductively derive classes of associations and larger systems of interacting elements within an area. To paraphrase Mabbutt (p. 11), "From the multiple wildlands attributes, their variations, and the intricacy of their interrelationships, it might be argued that each spot on the earth's surface is in the last degree unique and not to be compared with any other, but the scientific approach to wildlands classification involves the establishment of principles which

*This paper presents one phase of research conducted at the Jet Propulsion Laboratory, California Institute of Technology, under contract NAS7-100, sponsored by the National Aeronautics and Space Administration.

enable us to understand the nature of wildlands and compare them from place to place."

The analyses of natural environments have revealed the existence of recurring patterns formed by groupings of similar components, and to express this recurrence, the unit of description must be according to type, and not in terms of a set of regional entitites. Thus, Yapp developed the concept for vegetation mapping, wherein a characteristic set of plant species can be associated with specific terrain and soil conditions. Milne developed the catena concept for soils mapping, wherein complex soil associations could be considered as the repetition of a sequence of soil profiles in association with topography. Christian developed the concept of land units and land systems, as a recurring association series of vegetation, soils, and physiography, to achieve a mapping technique which would provide extensive predictive evaluation after a limited sampling effort. When applied in a quantitative analytical manner, as with the use of LANDSAT multispectral digital data, improvements in classification accuracy have been deserved when one incorporates the basic premise that recurring patterns can be derived inductively from the association of selected parameters obtained from sample sites.

The accuracy of the remote sensing classification is dependent upon the level of vegetation organization which is to be classified. A general classification for vegetation, proposed for International Biological Program (IBP) studies, recognized the following three levels: (1) formation, a general unit of different physiognomy; (2) region, including groups of communities with similar appearance and climatic controls; and (3) series, vegetation units with dominant climax species (Peterken, 1970).

Vegetation classification using LANDSAT has been approached from two basic directions: manual interpretation of images and computer-assisted automatic classification of digital data. Both manual procedures (Benson and Lauer, 1973; Kirby, 1973; Oswald, 1974; Heller et al., 1974), and computer procedures (Kirvida, 1973; Heath, 1974; Heller et al., 1974; Driscoll et al., 1974; Hoffer, 1975b, 1976) have been found to easily classify formation level vegetation with statistical accuracy in the 90 percent range. Accuracy is high because the lifeforms of formation level vegetation units (e.g., forest, grassland, shrubland) are sufficiently different to create contrasting spectral signatures on the LANDSAT image.

Considerably less accurate results have been reported for classifications at the regional level. Leachtenaeuer et al. (1973) and Heath and Parker (1973) found manual interpretation highly unsatisfactory, although Heller et al. (1974) and Lauer and Krumpe (1973) were able to obtain correct classifications in the 60 to 70 percent range. Computer classifications have been more successful, with Heath (1974), Heller et al. (1974) and Hoffer (1975b, 1976) achieving a somewhat better 74 percent (approximate) accuracy using Purdue University's LARSYS (Laboratory for Applications of Remote Sensing Systems) classifier. Driscoll et al. (1974) managed an average 95 percent with a similar automatic classifier at the regional level. This high accuracy, however, was produced using a simple classifiction scheme of only 5 classes, two of which were clouds and water.

At the series level, Oswald (1974) and Heller et al. (1974), judged conventional photo interpretation of LANDSAT imagery unacceptable, and Heller's computer techniques accomplished no better than a 48 percent accuracy. Driscoll et al. (1974) classified five species-specific forest cover units with an average accuracy of 64 percent, although the individual range varied from a low of 16 percent upwards. The general consensus is that manual interpretations of LANDSAT images are marginally successful, but automatic computer classification techniques are more promising (Heller et al., 1974; Kirvida, 1973).

At present, procedures for classification of series level forest cover or timber types using LANDSAT data have not been perfected with satisfactory accuracy. Weber et al. (1972), suggested that an accuracy level of 95 percent was necessary to meet forestry standards, although a correct interpretation level of 85 percent was determined by Anderson et al. (1976) to be the minimum acceptable accuracy for any classification based on remotely sensed data.

Improvements in classification accuracy can also be expected from incorporation of the nonspectral information into the classification algorithm. Many ecological and sylvicultural studies have shown the importance of the topographic parameters of slope angle, slope aspect, and relative elevation in determining vegetation composition (Strahler, 1972, 1975, 1977; Bowersox and Ward, 1972; Carmean, 1970; Despain, 1972; Hartung and Lloyd, 1969; Wikum and Wali, 1974). Incorporating these parameters into a classification scheme would significantly improve accuracy (Hoffer et al., 1975, 1975b, 1976; Oswald, 1974, Witmer, 1972).

Only two studies, thus far, have incorporated topographic information into a discrimination problem. Driscoll et al. (1974) determined spectral response variations of ponderosa pine stands at several slope angles and developed a linear model to describe the derived relationships. Simple classification of the one species type on three slope categories resulted in an average 57 percent accuracy. However, after the spectral signatures for each band were adjusted using the average covariance matrix for the medium slope category, classification performance increased to an average 79 percent. Driscoll concludes his work by stating, "Improvement in classification at the Series level by machine processing can be expected provided effects of slope steepness and aspect in relation to sun angle and sensor location are considered" (p. 1203).

Hoffer et al. (1975) registered elevation data to three selected Skylab-2 spectral bands and input those four channels to the LARSYS classifier. Improvements of 23 percent and 32 percent were obtained for two regional level forest cover types. While Hoffer et al. determined that "... the use of the elevation data offers considerable potential for improving the classification performance" can be achieved for individual forest cover types and concluded that ". . . a better analysis technique must be utilized to effectively incorporate the topographic data" (p. 27).

For the wildlands mapping projects described, the artificiality of essentially all resources classification schemes (ground based and otherwise) was tacitly accepted at the outset. Thus, during the initial phases of establishing the training statistics little concern was shown for trying to establish spectral clusters that corresponded to "known" resource types. Typal classifications of natural biological resources (vegetation, habitats, etc.) are most often only intuitive in substance and seldom adequately quantified. This condition has contributed to difficulties for testing LANDSAT classification as observed by Maxwell (1976) working with vegetation production in Colorado, Bentley et al. (1976) working with vegetation and soils in Arizona and Montana, and Hutchinson (1978) striving for an integrated survey method in the Mojave Desert of California. It appears that extreme variability within the ground control class areas commonly leads to improper evaluations of how well LANDSAT data can define resources.

## *KLAMATH FOREST STUDY*

### Introduction

The Klamath National Forest is located in northern California. This area includes 1,400 square miles of rugged terrain in the Siskiyou, Scott Bar, and Salmon Mountains. The topography ranges from 500 to 8,000 feet in elevation and includes very little developed area. The high relief and mountainous topography provides a diversity of habitat that is reflected in the forest cover. At lower elevations, oak and other hardwoods of little commercial value are dominant. Middle elevations are often characterized by stands of douglas fir, ponderosa and sugar pines. At the higher elevations, white and red fir are abundant. Slopes are moderate to steep over most of the area, and the forest cover typically varies considerably with slope aspect.

There is also significant climatic variation in Klamath National Forest. The western portion of the Forest receives abundant rainfall, and tends to be cooler than the eastern portion which is farther inland. There is also considerable geologic variation within the Forest. Volcanic activity has played an important role in landforms and development of the soils in the eastern portion of the Forest, producing extensive areas underlain by basalt flows. Tree species composition varies greatly between the two extreme ends of the Forest.

Previous research in the Doggett Creek watershed of the Klamath National Forest showed that LANDSAT information combined with digital terrain data could recognize species-specific forest cover classes with high accuracy (Strahler, Logan and Bryant, 1978). In this study, terrain data in the form of elevation and slope aspect was registered to a July LANDSAT frame. Combined with LANDSAT data, the registered terrain information was used in a supervised classification system in two different ways: (1) as pseudospectral channels in a conventional classification framework and (2) as conditioning variables to modify prior probabilities of occurrence of forest cover classes depending on observed aspect and elevation relationships. Accuracy, as evaluated by a separate random sample, was as high as 85 percent in distinguishing between such species-specific classes. The work at Doggett Creek showed that multispectral LANDSAT imagery is capable of distinguishing between forest stands containing trees of differing sizes and spacings, and that terrain data can provide information on the habitat preferences of species. This awareness led to the design of a two-track system described below.

## *SYSTEM OVERVIEW*

A two-track system was developed, in which one track processed LANDSAT data to yield timber volume information and a second track processed terrain data to yield forest composition information (Figure 1) (See Strahler, Logan and Woodcock, 1979). In the first track, LANDSAT data are clustered using an unsupervised classification algorithm to produce a large number of highly specific classes. In two editing phases this large number of classes is reduced to a much smaller number. The first is a statistical editing phase, in which classes of similar spectral characteristics are merged. The second editing phase is based on spatial relationships of classes, and merges classes with similar, contiguous occurrence patterns. In this way, a relatively small number of classes is obtained for which there is less variation in timber volume within clases than between classes. Once these volume-homogeneous classes are established, the mean timber volumes within them are determined by ground sampling using existing United States Forest Service cluster plot samples (USGS, 1976). These samples are processed by Forest Service software to yield estimates of timber volume within system-defined volume classes.

An important feature of the first-track techniques is the incorporation of a texture channel. This channel uses the standard deviation of Band 5 density values produced from a three-by-three pixel moving window as a variable to categorize spatial heterogeneity in the vicinity of a pixel. Use of this standard deviation texture data as an additional channel provides a differentiating factor which groups pixels together according to their degree of local uniformity in image tone.

The second track of the inventory system constructs an ecological model of species preferences for topographic positions based on ground sample data and reconnaissance mapping. Using elevation, slope angle, and slope aspect values for cluster plots as independent variables, a trend surface model is constructed which predicts the relative proportion of species to be found at a pixel as fitted to proportions calculated from the individual cluster plots.

The two tracks are merged at the pixel level (Figure 1). The pixel is classified, and a timber volume is associated with it. Registered terrain data then indexes an appropriate set of species proportions, and thus a timber volume increment for each species may be assigned to the pixel. Timber volume estimates for management areas, large and small, are then prepared by aggregating these volume-by-species estimates as required.

## *FOREST CLASSIFICATION*

The first track of the system involves classification of the pixels within a natural region into a small number of volume-homogeneous classes. This process begins with unsupervised clustering of the forested portions of the LANDSAT image and then proceeds through two phases of extensive editing of unsupervised classes.

A July LANDSAT frame (2540-18033 15 July 1976) with little apparent atmospheric degradation was selected for digital image processing. On this date the sun angle was high and shadowing was greatly reduced. Phenologically, both coniferous and deciduous signatures were well differentiated, as were signatures of brush, grass, and other ground covers likely to be interspersed with forest canopy.

### Clustering of the LANDSAT Image

The first step in establishing volume-homogenous classes is unsupervised clustering of the natural region. Because the LANDSAT image enclosing a natural region typically includes significant areas of nonforest land, it is more cost-effective to

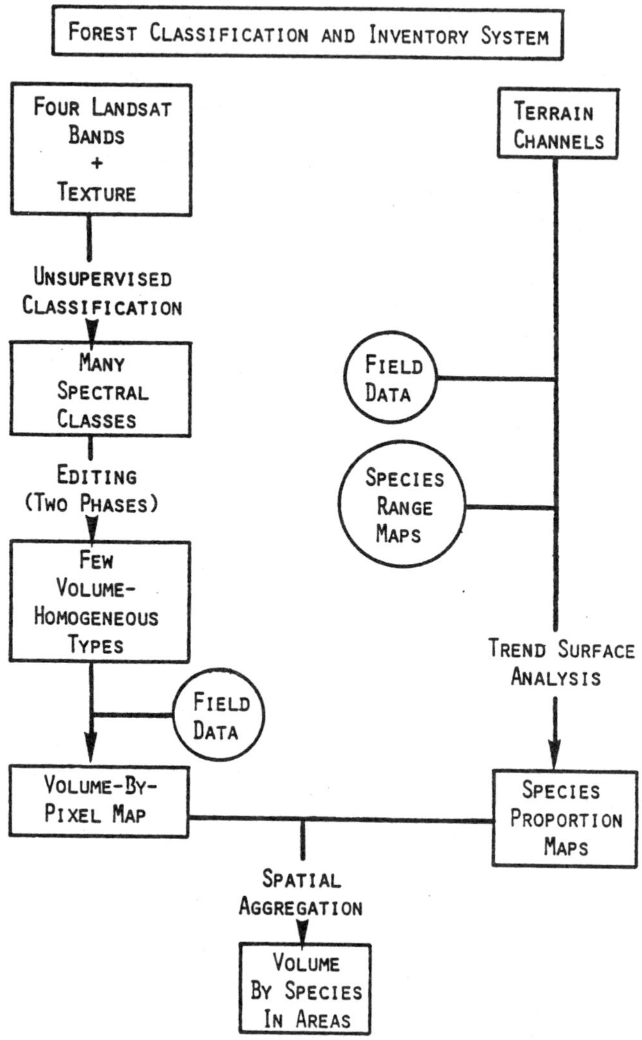


*Figure 1. Two-track outline of the forest classification and inventory approach.*

mask out all areas of the LANDSAT image not falling within established compartments before carrying out the unsupervised clustering.

Unsupervised clustering was accomplished by using the JPL-VICAR program USTATS, which incorporates a modified version of an algorithm obtained from Pennsylvania State University. This algorithm is based on a method suggested by Tyron and Bailey (1972) for clustering large numbers of observations. Beginning with the first pixel in the first scan line, each pixel is selected in turn. The Euclidean distance from the pixel to each existing cluster is calculated, and if it falls within a user-specified threshold, the pixel is merged into the cluster and the cluster centroid is recomputed. If the pixel does not fall within the distance threshold of any existing cluster, it becomes a new, single-pixel cluster.

When the clustering process is completed, clusters are ranked from largest to smallest, and user-specified number of centroids are output as classes for input to classification programs. The 200 largest classes were found to contain all the significant variation in the forested portion of the image, hence those 200 were retained for future processing. The output classes comprised a dataset which included (1) the number of pixels used to define each class; (2) the centroid (mean values) for each class; and (3) the variance-covariance matrix (in lower symmetric columnwise format) which characterized each class.

### Spectral Editing Phase

The USTATS algorithm produces an unwieldy number of tightly defined classes; these classes must be edited into a simpler, smaller set. This editing takes place in two phases. First, there is a spectral merging, based on the statistical similarity of classes. Second, there is a spatial-spectral editing, in which classes are combined depending on their spatial contiguity and spectral similarity. The spectral editing phase begins with processing of the 200 output classes using the NTSYS (Numerical Taxonomy SYStem) program package to "cluster the clusters." The NTSYS package was developed by Rohlf, Kisnpaugh, and Kirk (1974), at the State University of New York, Stonybrook, for use in numerical taxonomy. It allows the user to carry out many different clustering algorithms on the same dataset using a wide variety of resemblance measures.

For use in NTSYS, the class centroids are processed to calculate a symmetric matrix of standardized distances between all classes. The standardized distance is simply the Euclidean distance between classes calculated after standardizing each channel for the average within-class standard deviation. In this measure, small differences in values for channels which vary less (e.g., Bands 4 and 5) are equated with larger differences in channels which vary somewhat more (e.g., Bands 6, 7, and texture).

This standard distance matrix is then input to NTSYS, which constructs a dendrogram of the classes using a complete linkage algorithm. In this algorithm, a class or cluster of classes merges with the class or cluster to which it is least dissimilar. Experience has shown that the complete linkage algorithm produces the most satisfactory dendrograms, although other methods may produce trees which conform better to the structure of the original resemblance matrix.

The NTSYS dendrogram is used as a guide to condense the 200 classes, which are marked for merging into a smaller number of classes (perhaps 70-100). Merging is accomplished by a VICAR program which recalculates new classes of centroids by pooling old centroids as weighted by their size in pixels. New class variances are also computed for each channel. Small, nonforest classes may also be deleted using this program. The output classes are then inspected to assure that merging has not created any new classes with unduly high variances; if so, the merging program is rerun and such classes are split.

Generally, 70 to 100 classes remain after spectral editing based solely on statistical similarity. Further statistical merging only creates more poorly defined classes with increasingly large dispersion matrices. Thus, the image is now classified using this large number of classes. Although this classification is costly it is highly precise, subdividing the image into many, tightly defined units.

### Spatial-Spectral Editing Phase

The spatial-spectral editing phase of the classification process involves the aggregation of classes that are spatially contiguous as well as spectrally similar. This phase constitutes the final editing in the classification process, reducing the number of classes to approximately the same number of size- and density-differentiated strata used by the Forest Service. Depending on the diversity within the natural region in question, the final number of volume-homogeneous classes should range from six or seven in an area of low diversity to twelve to fifteen in an area of high diversity.

The first step in identifying classes to be aggregated is to cluster the classes remaining from the spectral editing phase, again using NTSYS. Although this clustering is based only on spectral similarity, experience has shown that many spectrally similar classes are also spatially contiguous. Thus, the NTSYS dendrogram is highly valuable as an editing guide.

Spatial contiguity of classes is established by interactive viewing of the classified image using the Image-100 at JPL. (Schaller and Towles, 1975, provide a description of this equipment.) The Image-100 allows the viewing of eight color-differentiated classes which can be overlaid on a LANDSAT background image. As classes are displayed (Figure 2), air photos of test compartments are inspected to verify that the classes to be aggregated contain trees of similar size and spacing. In addition, notes are prepared describing the geographical areas, timber types, and other general characteristics of the newly aggregated classes.

After a satisfactory volume-homogeneous strata classification is developed, a post-processing technique is applied to refine the classification through merger of isolated pixels with the surrounding areas. The technique (VICAR program SIMPLIFY) reduces the spatial heterogeneity characteristics of digital classification, producing a product more appropriate for Forest Service practices which are based on five-acre minimum units (Figure 2). The algorithm reclassifies a pixel if a user-defined number of pixels (typically four) are all of the same class in a surrounding window (typically three-by-three). Final classifications were simplified by repeating this method twice (Figure 2).

### Associating Volume With Class

After the final classes or strata have been identified in the spatial editing phase, the actual per-acre timber volume associated with each class must be determined. The methods used in assigning volumes to the LANDSAT derived classes are closely tied to those used by the Forest Service, so that final inventory estimates will be directly comparable with existing estimates. A brief summary of the Forest Service methodology follows below.

In the Klamath National Forest as well as many others, ground data are collected in the form of "cluster plots." Termed cluster plot because each plot includes a cluster of five subplots, these timber volume samples are located according to a stratified random sampling plan. In conventional inventory procedures, the strata are defined using stand maps produced from air photos. Four to six cluster plots are selected for each stratum present in a compartment. The data from the cluster plots are processed by a computer program at Forest Service Laboratories in Fort Collins, Colorado, to produce the volume estimates later used for timber management.

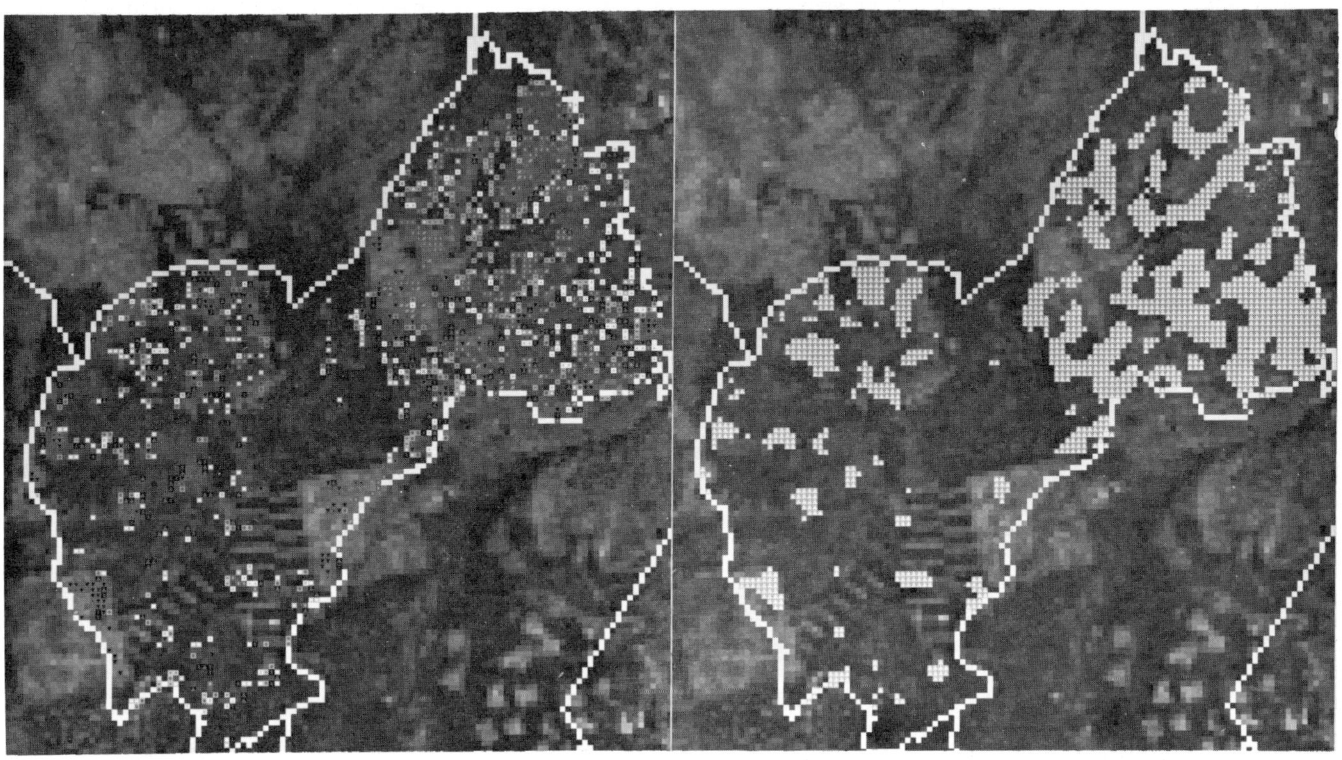

*Figure 2. Spatial-spectral editing of Eagle compartment. The twelve classes shown in the left image were combined on the basis of their spatial contiguity and spectral similarity. The right image shows the final appearance of the class after simplification, which removed isolated pixels and smooths boundaries. Although less spatially accurate, the simplified image is often better suited to management use.*

These same cluster plots are used for determining the mean timber volume of the LANDSAT derived strata. The cluster plots are located on the LANDSAT image by transforming their UTM (Universal Transverse Mercator) coordinates into LANDSAT line and sample coordinates. They are then aggregated according to membership in the LANDSAT defined strata, and the cluster plots are processed using conventional Forest Service software. The use of Forest Service data and processing assures uniformity of assumptions concerning factors influencing volume estimation, such as the degree of taper due to species and the reduction in volume due to conks, rot, cull, breakage, and other defects.

## *MODELING SPECIES PROPORTIONS THROUGH ECOLOGICAL RELATIONSHIPS*

The second track (Figure 1) uses ecological relationships based on topographic data to predict the proportions of species which will occur at a pixel. Past work in the Doggett Creek area showed that the incorporation of terrain information into a classification system greatly increases the accuracy of identification of species-specific forest cover units (Strahler, Logan, and Bryant, 1978). After working with LANDSAT and topographical data for several months, it became apparent that the terrain information was separating species based on their habitat preferences.

For example, the general altitudinal zonation observed at Doggett Creek showed red fir present in highest elevations, with zones dominated by white fir, douglas fir, and ponderosa pine ranging respectively to the lowest elevations at the Klamath River. Although these closed canopy, coniferous classes were spectrally similar, the use of elevation in the classification procedures as a pseudospectral channel allowed separation of species-specific classes. This work provided the basis for the two-track concept, in which LANDSAT data are used to estimate the size and spacing of forest tree crowns, whereas topographic position is used to predict the species proportions through trend surface analysis.

At the present time, final trend surface models are not complete for any natural region. However, a pilot study modeling the species proportions for the Walker Creek compartment, in the Oak Knoll Ranger District, has been carried out. It confirms the validity of the concept. This study is discussed in more detail in a later section.

### Digital Terrain Data Processing

The terrain data used consist of Defense Mapping Agency (DMA) digital terrain tapes, now distributed by the U.S. Geological Survey through the National Cartographic Information Center at Reston, Virginia. These data were created by scanning 1:250,000 scale topographic maps for elevation values along contour, stream and ridge lines, and at point locations. These values are processed to produce a grid of elevations for every 0.01 in. on the map, or approximately 200 feet on the ground. Because the digital terrain data are prepared from 1:250,000 maps, the landforms shown are quite generalized and the fine grid with which the values are recorded often implies a degree of precision not inherent in the original information.

Registration of the terrain image to the LANDSAT scene takes place using a resampling algorithm which uses a two-dimensional correction grid derived from control points to convert each pair of LANDSAT coordinates to digital terrain coordinates. These calculated coordinates are then used to select the terrain pixel (or an interpolated value of the pixel) to appear in the registered output image. Selection of control points which identify the spatial correspondence between digital and LANDSAT images can be quite difficult, since terrain features are not necessarily reflected in the LANDSAT image. Mountain peaks and river intersections provide the most reliable locations for registration control.

With the terrain data registered to the LANDSAT scene (Figure 3), separate images of slope angle and slope aspect can be generated from elevation values. These values are generated by the least squares fitting of a plane through each pixel and its four nearest neighbors. The output slope angle is scaled to reflect 0-90 degrees within the range 0-255, and slope aspect is scaled so that the aximuth in degrees (0-359) also varies from 0-255. The aspect image is then transformed using a cosine function with a shifted axis, following the suggestion of Hartung and Lloyd (1969). In this transfomration, zero represents northeast, 128 represents southeast or northwest, and 255 represents southwest. This function is based on the ecological observation that sites on northeast-facing slopes are most productive, those on southwest facing slopes are least productive, and those on southeast and northwest slopes are intermediate in character (Figure 3).

### Walker Creek Modeling

As a pilot test of the proportion-modeling concept, existing data recording basal area for commercial forest species in the Walker Creek compartment of the Oak Knoll Ranger District were processed. Inventory cluster plot data were obtained for this compartment from the Forest Service, and additional plots were taken as needed to assure six samples each at nine possible combinations of low, middle and high elevations and southwest, northeast, and neutral aspects. Basal areas were calculated, converted to proportions, and transformed by arcsine root. A two-way analysis of variance was performed for each species, and the treatment effects were used to define a trend surface constructed of eight intersecting planes connecting the nine two-way treatment effects in measurement space. This surface was used in predicting species proportions for the entire basin on a pixel-by-pixel basis.

Figure 3 presents the topography of the compartment as well as the predicted proportions for douglas fir, white fir, ponderosa pine, and black oak. The first pair shows the dominance of white fir over douglas fir at higher elevations as well as the preference of douglas fir for southwest-facing slopes. The latter pair shows the preference of both ponderosa pine and black oak for lower elevations, in addition to their marked, but different, aspect preferences. Although these predictions have not yet been confirmed by ground sampling, they are quite consistent with general knowledge of the area. All evidence from this test area confirms the appropriateness of the topographic model for use in forest classification and inventory according to species.

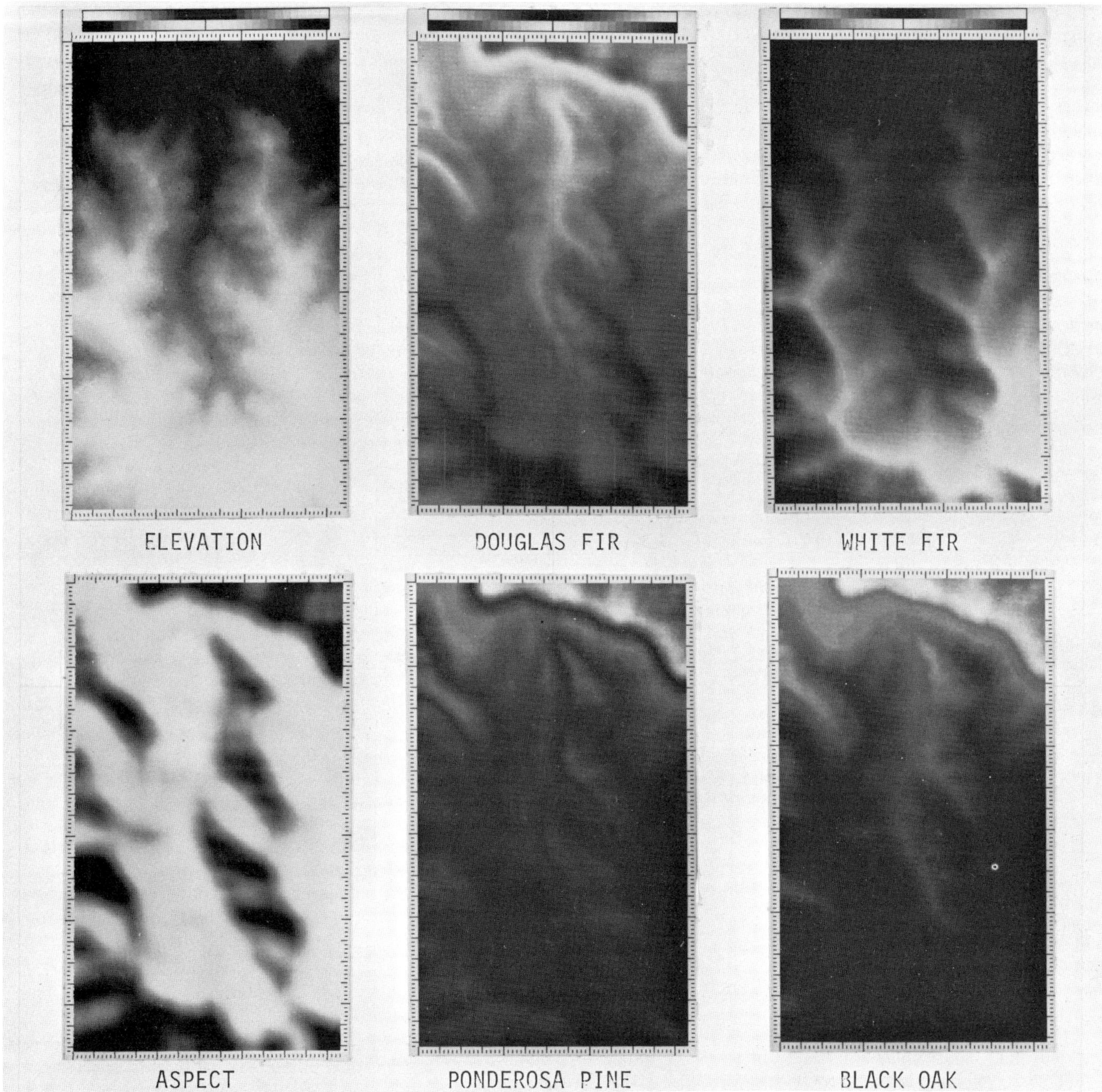


*Figure 3. Elevation and aspect images of Walker Creek vicinity with predicted basal area maps of four forest species. Elevation, slope, aspect and USFS Cluster Polt data were used to construct a trend surface model that predicts the relative proportions of species to be found at a pixel (Images are stretched from black to white representing proportions as follows: douglas fir, .07 - .86; white fir, 0 - .71; ponderosa pine, 0 - .25; black oak, 0 - .30).*

## *CALIFORNIA DESERTS STUDY*

### Introduction

The Bureau of Land Management (BLM) in the State of California has been mandated by Congress to prepare a comprehensive, multiple use management plan for the California Desert Conservation Area (CDCA) by October 1, 1980 (Federal Land Policy and Management Act, Section 601, 1976). The CDCA encompasses an area of approximately 25 million acres, or about one-fourth of the State of California. Resource information for this sparsely occupied portion of the state is incomplete in many areas and nonexistent in others. Time and budget constraints have precluded the use of conventional mapping and inventory techniques for obtaining much of the needed resource data. The feasibility of using multispectral classification of LANDSAT imagery for the CDCA Plan was explored by representatives of the BLM Desert Plan Staff at the EROS Data Center in early 1977. In general the results were highly favorable and the conceived potential was found to greatly exceed the initial expectations. However, different systems and procedures needed to be implemented if an approach to uniform classification over the entire region was to be realized (Bryant et al., 1979).

## LANDSAT MOSAIC DATA BASE

Considerable work has been done at JPL's Image Processing Laboratory developing geometric rectification and registration algorithms, and more recently, software has been developed for digital image map reprojection and image mosaicking.

In the case of LANDSAT mosaicking, it was necessary to incorporate ground control points of known position in the image, as well as relative registration points to fine-tune the local misregistration between frames (Zobrist, 1978). The reasons for incorporating known ground control points are two fold: a) the MSS multispectral scanner on LANDSAT is not a framing imaging system, so that continuous changes in pointing perspective geometry make it virtually impossible to reconstruct a perfect orthophoto image from calibration data; and b) the relative position of points on the earth's surface is precisely known, with the result that geodetic control points must be used as input to the geometric correction of the satellite image data if any satellite mosaic is to be expected to conform to the planimetry of existing maps. All the imagery used to construct the CDCA mosaic data base was, therefore, resampled in the x and y directions to accommodate LANDSAT sensor and orbital characteristics, local topographic offset effects, and to conform to the Lambert Conic Conformal Map Projection. The pixel resolution selected was 80 meters square. The data were also resampled in the z domain to present a smooth surface that no longer reflected abrupt changes in sensor calibration or day-to-day side-lap differences associated with atmospheric effects.

The images used to generate the CDCA regional mosaic were ten LANDSAT scenes taken on four sequential cloud-free days in August, 1976. A detailed analysis of the images generated has shown that the brightness value adjustment (z correction) has successfully smoothed out any sharp edge effects between frames due to atmospheric diferences at different days, thereby permitting the extension of thematic classification across LANDSAT frames (Zobrist & Bryant, 1979). Converting the image data to a specific map projection allows for easy, repeatable access to any portion of the mosaic. Coordinates of the vertices of an area requiring access are specified in latitude and longitude (degrees, minutes, seconds) which are in turn converted into the x and y positions of the image raster grid. This capability is useful for detailed inventory and analysis of sub areas, since the CDCA mosaic comprises some 50 million picture elements (pixels) per LANDSAT band (Figure 4). Such a corrected image grid permitted the pixel for pixel registration of the National Cartographic Information Center/Defense Mapping Agency (NCIC/DMA) digital terrain image data base to the mosaicked LANDSAT imagery and its use as an aid in thematic classification.

## OBTAINING CLASSIFICATION STATISTICS

Despite the fact that the mosaicking procedure's brightness value adjustment (z correction) had successfully smoothed out any sharp edge effects between frames due to atmospheric differences on different days, it was felt that assumptions of signature extension from selected ground plots across LANDSAT scene boundaries should probably not be made. Even though desert regions are characterized by stable atmospheric conditions for long periods of time, changes in plant species composition for similar habitats are known to develop over several degrees of latitude, as are reflectivity signatures for similar areas. A system of linear transect sampling was adopted as a means to assure uniform classification statistics with sufficient discrete classes to accommodate essentially all resource cover types defined by the BLM resource specialist. The initial pixel brightness values for all four MSS bands were extracted in two pixel wide swaths along approximately 3,000 miles of transects specified by the Bureau of Land Management science team to include all variable types in the landscape (see Figure 5). The transects traversed all types of terrain elevation, slope gradient, slope aspect, and the various cultivated and populated areas. The transect data were then aggregated into one small image per LANDSAT band which represented one-half of one percent (0.5%) of the entire CDCA. An unsupervised clustering algorithm was applied, which initially separated out 1993 clusters.

The unsupervised clustering algorithm, which builds up clusters as it passes through the training data was structured to maximize the number of initial clusters. The initial cluster data set displayed the very broad range in reflectivity values found in the California desert regions, which range from highly reflective white sands and playas to dark basalts, and irrigated agriculture to forested mountain areas, as well as the diverse variety of rangeland and dry cropland conditions. Much of the CDCA is sparsely vegetated, i.e. has less than 10% plant ground cover, and maximum ground cover seldom exceeds 50% except in cultivated lands which were of little concern to BLM scientists. Wildland areas, with the higher percentage ground cover values, though comparatively small in size, are of great significance in terms of total plant productivity and related resource values. The task of reducing the 1993 clusters to a manageable number became a problem of devising a way to retain significant but small clusters having strong vegetation components in their signatures. It is generally known that such clusters should be those with comparatively low reflectance in the red band (MSS 5) and high reflectance in the infrared bands (MSS 6 & 7) (Maxwell, Bently, et al). The BLM science team developed a systematic procedure to reduce the initial 1993 clusters into 100 principal clusters while maintaining the critical elements needed to define landscape diversity. To that end, the 1993 initial clusters were first separated into two partitions based upon the strength of the vegetation component of the spectral signature as indicated by the brightness ratio of MSS band 6 to MSS band 5. If the ratio of MSS band 6 to MSS band 5 was greater than or equal to one, the clusters were determined to signify a chlorophyll type signature and if the ratio was less than one, the clusters determined a nonchlorophyll signature. This basically broke the clusters into vegetation reflecting (684 clusters) and non-vegetation reflecting signatures (1309 clusters). To further reduce the

*Figure 4. Ten frame LANDSAT mosaic encompassing the California Desert Conservation Area.*

amount of clusters each resultant group was broken down into two partitions. Of the vegetation reflecting signatures where MSS band 6 was greater than MSS band 5, these signatures were divided into groups where MSS band 7 was greater than or equal to MSS band 5 (very high vegetation signature content; 363 clusters) and where MSS band 7 was less than MSS band 5 (high vegetation signature; 321 clusters). The non- or low vegetation clusters were divided on the basis of whether MSS band 5 brightness was greater than 130 (on a pixel reflectance scale of 0-255), (very bright surface conditions, e.g. playas and sand; 690 clusters) and whether MSS band 5 brightness was less than or equal to 130 (dark surface features, e.g. basalt flows, shadows, 619 clusters). This resulted in partitioning the 1993 initial clusters in four groups representative of four major signatures types in the CDCA.

The reduction of the clusters to approximately 25 classes within each of the four partitions was performed by first ranking the clusters in accordance with their brightness coefficient (Figure 6), and then by merging overlapping clusters and deleting the population and associated statistics of clusters

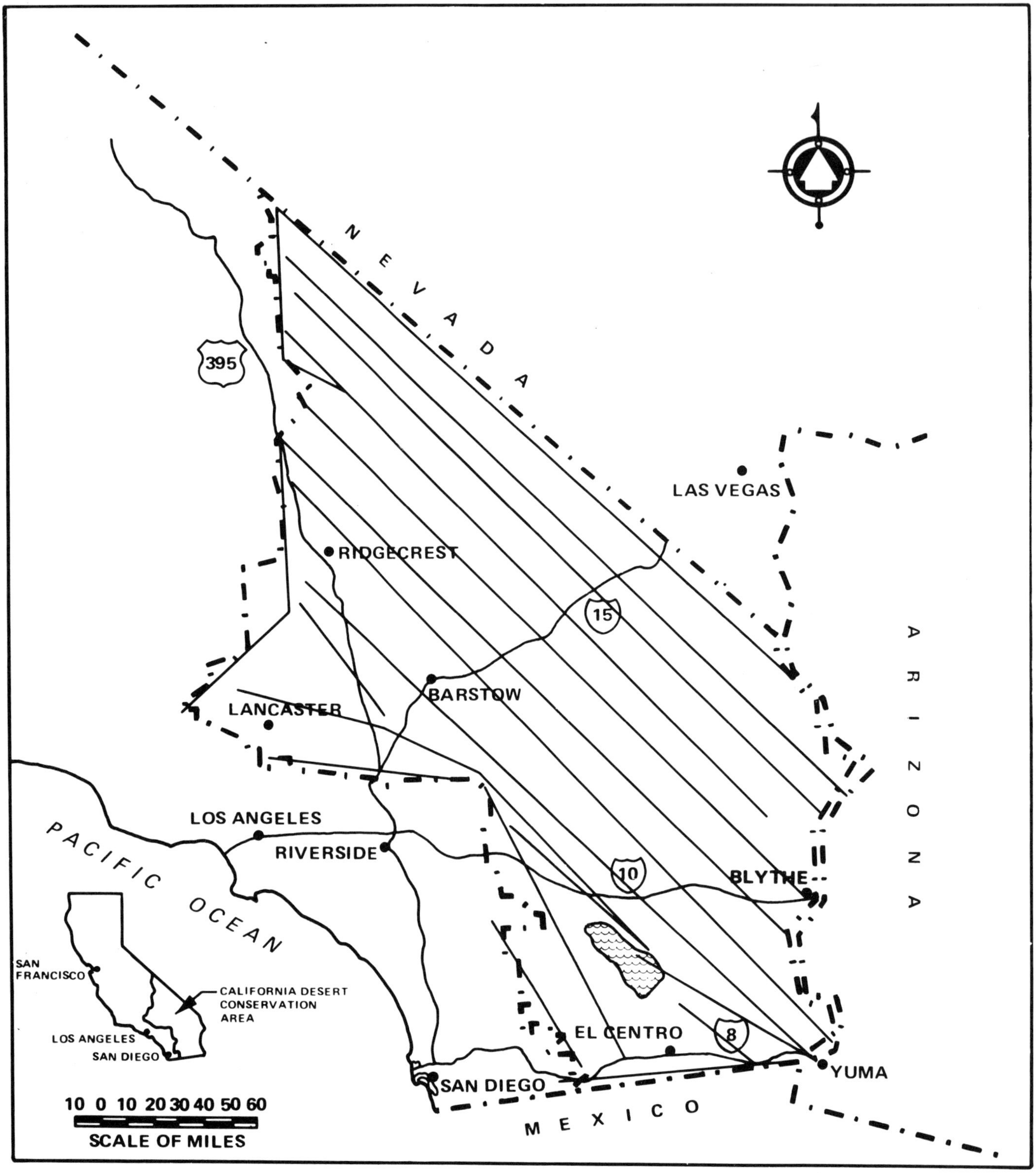


*Figure 5. Location of Transect Lines Used to Gather Training Sample Statistics for California Desert Conservation Area Multispectral Classification.*

1. For Clusters 1-1993 $\frac{B_4 + B_5 + B_6 + B_7}{4} = X$

where $B_4$-$B_7$ are Landsat Bands 4-7.

2. Sort 4 in ascending order to find minimum

Minimum Aggregate Cluster $= \sum X_{MIN_{1-4}}$

$$CL_{MIN} = \underset{B_4}{MIN\ 1},\ \underset{B_5}{MIN\ 2},\ \underset{B_6}{MIN\ 3},\ \underset{B_7}{MIN\ 4}$$

3. For clusters 1-1993 compute:

$$B_4 - MIN_1 = M1$$
$$B_5 - MIN_2 = M2$$
$$B_6 - MIN_3 = M3$$
$$B_7 - MIN_4 = M4$$
$$\sum MIN_{1-4} \sum M_{1-4}$$

4. Brightness coefficient for each cluster:

$$CL = \frac{\left(\sum M_{1-4}\right)^{2}}{\left(\sum M_{1-4}\right) + \left(\sum MIN_{1-4}\right)}$$

*Figure 6. Procedure Used to Determine Brightness Coefficient (after Meuller-Dombois & Ellenberg (10)).*

where the mean value occurred within the one standard deviation ellipse of a cluster with a larger population. In order to assure that population size and standard deviation about the mean value of a cluster did not predominate in the cluster merging process, each partitioned group was treated individually. Important low frequency pixel clusters representing strong vegetation signatures were thus retained to be used in the classification process (Table I).

## CLASSIFICATION

The final one hundred clusters selected for the CDCA were then used as input to a hybrid Bayesian/Parallelepiped type classifier and applied to four primary test areas comprising over 400,000 acres each (Addington, 1975). These areas were 25.4 miles (40.9 km) on a side or 512x512 pixels. Eight secondary test areas were also used for classification verification. These test areas were selected by the BLM Desert Plan science team for their diversity and representativeness of the varieties of landscape found throughout the CDCA. The four primary test areas were intensively studied and analyzed by the BLM staff in their office by using black and white film overlays of each class on color transparencies of the raw LANDSAT data. As general patterns of class distributions emerged, names were given to the classes. Several class combinations for both primary and secondary test areas were then displayed on an interactive display system at JPL, and analyzed by the BLM science team to permit the assignment of each class or

Table 1

*RELATIVE PERCENT OF PIXELS AND CLUSTERS IN THE BAND RATIO-BRIGHTNESS CATEGORIES FROM THE TRANSECT SAMPLE TRAINING SET*

| | LANDSAT Band Ratio-Brightness Categories | | | |
|---|---|---|---|---|
| | 1<br>Band 5 ≤ Band 6<br>Band 5 ≤ Band 7 | 2<br>Band 5 ≤ Band 6<br>Band 5 > Band 7 | 3<br>Band 5 > Band 6<br>Band 5 ≤ 130 | 4<br>Band 5 > Band 6<br>Band 5 > 130 |
| Percent of Pixels | 6% | 12% | 52% | 30% |
| Percent of Clusters | 18% | 16% | 35% | 31% |

class combination to either a plant community, or a soil series. Class assignment was aided by extensive supplemental ground truth collected by the BLM staff through field surveys and low altitude aerial photography. The ground truth information helped spot inconsistencies in classes from test area to test area. It was found, for instance, that with the aid of registered NCIC/DMA elevation data, and derived slope aspect data, many of the inconsistencies in class assignments were resolved. Changes in elevation stratification of communities with similar reflectance characteristics often occured between test areas in the high desert and those in low desert areas. In a similar manner, problems such as dark desert pavement and shaded north slopes in mountains could be differentiated.

Once the BLM science team was assured of the proper class assignments to soils, and vegetation types, the respective maps were generated for each of the 22 quadrangular areas encompassing one degree of latitude and longitude in the CDCA. As results of the 12 test areas analyses showed a systematic drift in signature characteristics from north to south, the quadrangles were each assigned to one of five quadrangle groups. Each quadrangle was considered as an individual unit for further analysis and resource inventorying. During the analysis, resource classes were to be inventoried and aggregated by several administrative districts for tabulations. The first of these districts are leased grazing allotments in the California deserts. The grazing allotments are digitized from 1:250,000 scale maps in the 1° of latitude by 1° of longitude format in order to be consistent with the LANDSAT quadrangle areas. In addition to grazing allotments, a file of public and private lands is digitized in the same manner in order to aggregate resource classes in grazing allotments by land ownership. The resource data for the area is "packaged" into manageable sizes for more detailed investigation during the comparative study phase needed to draw up a comprehensive plan for use of the CDCA. Furthermore, the 1° x 1° quadrangle data sets can be easily accessed on BLM minicomputer systems in the future.

## *ACKNOWLEDGMENTS*

The authors would like to thank Dr. Alan Strahler, Dr. Nevin Bryant, Curtis Woodcock, and Al Zobrist for their significant and valuable contributions to this effort.

## REFERENCES


Addington, J. D., "A Hybrid Classifier Using the Parallelepiped and Bayesian Techniques," *Proceedings* of the American Society of Photogrammetry, 41st Annual Meeting, March 1975, pp. 772-784.

Benson, A.S., and D. T. Lauer, 1973. "Testing Multiband ERTS-1 Imagery for Classification of Commercial Conifer Forests," in: *Aerial Color Photography in the Plant Sciences and Related Fields, Proceedings.* Fourth Biennial Workshop Held at the University of Maine, Orono, July 10-12, 1973. College of Agriculture and Life Sciences, Maine University, Orono, with American Society of Photogrammetry.

Bently Jr., Gordon, Bette C. Salmon-Drexler, William J. Bonner and Robert K. Vincent, *Landsat study of ephemoral and perennial rangeland vegetation and soils, United States Department of Interior,* Bureau of Land Management Report sponsored by National Aeronautics and Space Administration under contract number S-53766-A, December, 1976.

Bryant, N. A., H. B. Johnson, R. G. McLeod, and A. L. Zobrist, (1979). "California Desert Resource Inventory Using Multispectral Classification of Digitally Mosaicked Landsat Frames," *Proceedings,* Symposium on Machine Processing of Remotely Sensed Data, Purdue University/LARS, West Lafayette, IN, June 27-29, 1979.

Christian, C. S. (1958), "The Concept of Land Units and Land Systems," *Proceedings* 9th Pacific Science Congress, 1957. 20:74-81.

Driscoll, R. S., E. R. Francis, J. A. Smith and R. A. Mead, 1974. "ERTS-1 Data for Classifying Native Plant Communities — Central Colorado," in: *Proceedings Ninth International Symposium on Remote Sensing of Environment,* April 1974, Vol. 2. Center for Remote Sensing Information and Analysis, Environmental Research Institute of Michigan, Ann Arbor.

Hartung, R. E., and W. J. Lloyd, 1969, Influence of aspect on forests of the Clarksville Soils in Dent County, Missouri: *J. Forestry,* Vol. 67, pp. 178-182.

Heath, G. R., 1974. "ERTS data tested for Forestry Applications," *Photogrammetric Engineering.* 40(9): 1087-1091.

Heller, R. C., R. C. Aldrich, R. S. Driscoll, R. E. Francis, and F. P. Wever, 1974. "Evaluations of ERTS-1 data for Inventory of Forest and Rangeland and Detection of Forest Stress," NASA-ERTS SA-ERTS Report #75-10147.

Hoffer, R., and staff of LARS, 1975. *Computer-Aided Analysis of SKYLAB Multispectral Scanner Data in Mountainous Terrain for Land Use, Forestry, Water Resources, and Geologic Applications, LARS Information Note 121275.* Laboratory for Applications of Remote Sensing (LARS), Purdue University, West Lafayette, IN.

Hoffer, R. M. and staff, 1975b. *Natural Resource Mapping in Mountainous Terrain by Computer Analysis of ERTS-1 Satellite Data,* Research Bulletion 919 (LARS Information Note 061575). Laboratory for Applictions of Remote Sensing and the Agricultural Experiment Station, Purdue University, West Lafayette, IN.

Hoffer, R. M., 1976. *Techniques and Applications for Computer-Aided Analysis of Multispectral Scanner Data,* LARS Information Note 062276. Laboratory for Applications of Remote Sensing (LARS), Purdue University, West Lafayette, IN.

Hutchinson, Charles F., "Digital Use of Landsat Data for Integrated Land Resource Survey: A Study in the Eastern Mojave Desert," Ph.D. Dissertation, University of California, Riverside, 1978.

Kirvida, L., 1973. "Computerized Interpretation of ERTS Data for Forest Management," in: *Symposium on Machine Processing of Remotely Sensed Data,* LARS, Purdue University, West Lafayette, IN, 1973.

Lauer, D. T., and P. F. Krumpe, 1973, "Testing and Usefulness of ERTS-1 Imagery for Inventorying Wildland Resources in Northern California," in: Symposium on Significant Results Obtained from the Earth Resources Technology Satellite-1, March 5-9, 1973, Vol.: Technical Presentations. NASA Goddard Space Flight Center, MD, 97-104.

Leachtenauer, J. C., C. E. Elworth, and R. A. Schindler, 1973. "Image Analysis Techniques for Timber Mapping," in: Proceedings 4th Annual Conference of Remote Sensing in Arid Lands. University of Arizona, Tucson, College of Earth Sciences, Office of Arid Land Studies, 3140328.

Mabbutt, J. A. (1968), "Review of Concepts of Land Classification," in G. A. Stewart ed., Land Evaluation; Papers of a CSIRO Symposium Organized in Cooperation with VNESCO 26-31 August, 1968, Melbourne, Australia, McMillan, p. 11-28.

Maxwell, Eugene L., "A Remote Rangeland Analysis System," *Journal of Range Management* 1976, Vol. 29, pp. 66-77.

Meuller-Domois, Dieter and Heinz Ellenburg, *Aims and Methods in Vegetation Ecology,* John Wiley & Sons, New York, NY 1974, pp. 212-217.

Milne, G., (1935), "Composite Units for the Mapping of Complex Soil Associations," *Transactions,* Third International Congress on Soil Science, Oxford, England, 1:345-347.

Oswald, E. T., 1974, "Evaluation of ERTS Imagery for Vegetation Interpretation in British Columbia," *Canadian Surveyor,* 28 (2): 147-154.

Peterken, G. F., 1970, "Guide to the Check Sheet for International Biological Program Areas Including a Classification of Vegetation for General Purposes," *International Biological Program Handbook 4.*

Schaller, E. S. and R. W. Towles, 1975. Image-100 — The interactive mutispectral image procesing system: *Proc. NASA Earth Resources Surveying Symp.,* vol. I-B, Houston, TX, pp. 1275-1292.

Strahler, A. H., T. L. Logan, and N. A. Bryant, 1978. Improving forest cover classification accuracy by incorporating topographic information: *Proc. Twelfth Int. Symp. on Remote Sensing of the Environment,* Manila, Philippines, April 20-26, 1978. Environmental Research Institute of Michigan, Ann Arbor, MI, pp. 927-941.

Strahler, A. H., 1972. "Forests of the Fairfax Line, Virginia and West Virginia," *Annals of the Association of American Geographers,* 62: 664-684.

Strahler, A. H., 1975. "Continuous and Binary (Presence or Absence) Measurement of Vegetation: A Comparison for Eastern Forest Data," *Proceedings: Association of American Geographers.* 7: 233-237.

Strahler, A. H., 1977. "Response of Woody Species to Site Factors in Maryland, U.S.A.: Evaluation of Sampling Plans of Continuous and Binary Measurement Techniques," *Vegetatio,* 35: 1-19.

Strahler, A. H., T. L. Logan, C. E. Woodcock, (1970) "Forest Classification and Inventory System Using Landsat, Digital Terrain, and Ground Sample Data," *Proceedings* 13th International Symposium on Remote Sensing of the Environment, April, 1979, Ann Arbor, Michigan.

Rohlf, F. J., J. Kishpaugh, and D. Kirk, 1974. *NT-SYS: Numerical Taxonomy System of Multivariate Statistical Programs.* New York State University, Stony Brook, and Quantra Corporation.

Tryon, R., and D. Bailey, 1972. *Cluster Analysis.* McGraw-Hill; New York, NY, pp. 147-150.

Yapp, R. H. (1922). "The Concept of Habitat," *Journal of Ecology,* 10:1-17.

Zobrist, A. L., "Multiple-Frame Full-Resolution Landsat Mosaicking to Standard Map Projections," *Proceedings* American Society of Photogrammetry Fall Technical Meeting, Albuquerque, NM, October 15-20, 1978, pp. 608-616.

Zobrist, A. L. and Bryant, N. A., "Map Characteristics of Landsat Mosaics," *Proceedings* American Society of Photogrammetry Annual Convention, Washington, DC, March 19-23, 1979, pp. 260-273.

# *Regional Environmental Studies: Application of the Geoecology Data Base*

*by R. J. Olson, J. M. Klopatek, and C. J. Emerson**

## *INTRODUCTION*

Regional studies examine the spatial, temporal, and functional characteristics of geographic areas whose size and boundary configuration often depend on the definition of a particular problem. Thus, reference may be made to ecological, urban, economic, political, or abstract regions. To address environmental problems on a regional scale, the county-level Geoecology Data Base was developed at Oak Ridge National Laboratory (ORNL). Using selected variables from existing data sources, a standardized digital data base was created for the conterminous United States. Over 1000 variables on file for county-subcounty units provide data on terrain, water resources, forestry, vegetation, wildlife, agriculture, land use, climate, air quality, population, and energy. This paper describes the data base and illustrates the utility of the county-level data base with three examples.

The Geoecology Data Base is utilized in conjunction with statistical analysis (SAS) and cartographic display (SYMAP, EZMAP, etc.) programs both to characterize regional environmental systems and to delineate possible environmental impacts. Much of the effort involved in the creation and expansion of the Geoecology Data Base consists of obtaining and editing diverse inventory and monitoring files. An integral aspect of this activity is the promotion of data exchange through increased awareness of available data and development of more efficient exchange methods. Data-base inventories and exchange standards are described as a joint product of the eight Department of Energy (DOE)-funded laboratories.

The three examples involve regional environmental studies conducted by the Environmental Sciences Division (ESD) at ORNL. First, future patterns of projected levels of $SO_2$ are overlaid on $SO_2$-sensitive agricultural crops and forest resources. Second, those areas of potential natural vegetation of the United States as modified or replaced by current land use are depicted. Finally, an ecological evaluation of potential wilderness areas using a number of county-subcounty ecological parameters is discussed. These applications demonstrate the versatility of county-level digital files, such as the Geoecology Data Base, in conducting regional environmental studies.

## *GEOECOLOGY DATA BASE*

The Geoecology Data Base was created by selecting variables from extant sources and storing them in a standardized, county-level file (Olson and Goff 1976). The County-Level Integrated Data Base (CLIDE) used within the CEQ UPGRADE system also uses this approach (CEW 1978). The emphasis has been on applying the data base to ongoing research needs through collaboration with specific projects. However, the data base can be accessed directly by investigators. Expanding and updating of the data base is also done according to project needs and funding within the Environmental Sciences Division.

### Data Base Design and Contents

The Geoecology Data Base was designed to meet the needs of regional environmental studies while accomodating the format of the majority of available environmental data. The data base method of easily overlaying thematic data stored at a common spatial unit (counties) for use in regional models was established during the NSF-funded International Biological Program (Figure 1. Burgess and Kern 1973). A prime consideration was to have a flexible, user-oriented system which could be developed, utilized, and upgraded with minimal costs.

Regional studies require data at various spatial scales (cell sizes) and for different geographic regions. Counties were selected as the primary geographic cell. They represent well defined spatial units which can be selected or aggregated based on various regional definitions. Crop-reporting districts, state climatic divisions, land resource regions, air quality control regions (AQCRs), water subareas, standard metropolitan statistical areas (SMSAs), Bureau of Economic Analysis areas (BEAs), and states all represent areas defined in terms of aggregates of county units. Furthermore, counties or regional aggregates of them are often the political units for implementing policy. In addition, a large amount of environmental data is

*Research sponsored by the Office of Technology Impacts, U.S. Department of Energy, under contract W-7405-eng-26 with Union Carbide Corporation. Publication No. 1389, Environmental Sciences Division, ORNL. Presented at: Harvard Computer Graphics Week '79, July 15-20, 1979, Cambridge, Massachusetts.

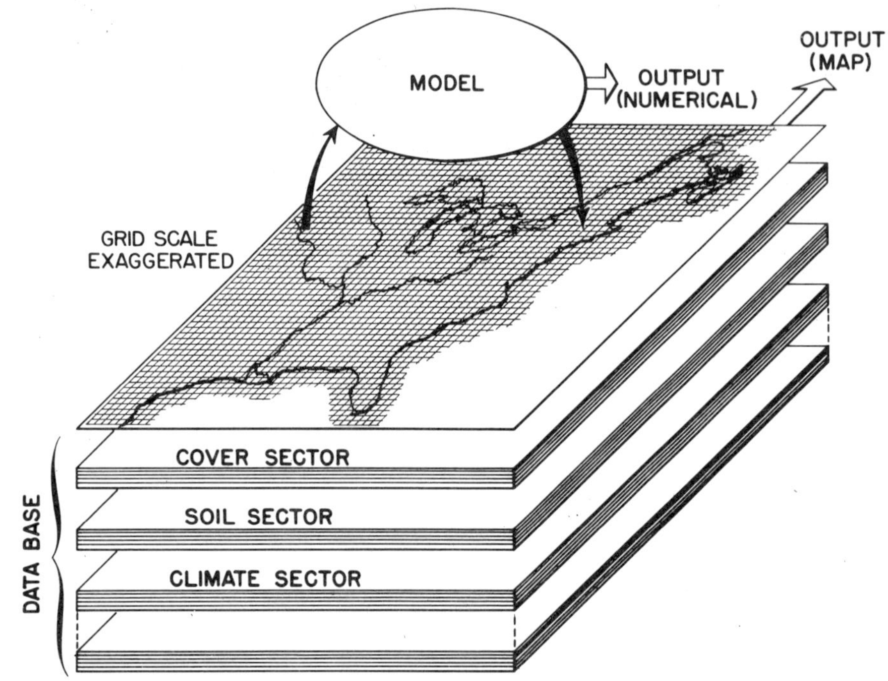


*Figure 1. Geoecology Data Base Design Concept (Burgess and Kern 1973).*

available as county statistics. Large (and often ecologically more heteroegeneous) counties are less desirable as spatial cells for regional analysis. Subcounty units, based on physiographic considerations, have been designated for the larger counties in the eastern states (e.g., Florida, Maine, Minnesota) in order to store data in more environmentally homogeneous cells. Yet to be done is subdivision of the large western counties. Independent cities, such as those found in Virginia, have been combined with appropriate county units as environmental data are normally not available for these smaller, urban areas.

Most data sets within the data base provide geographic coverage of, or are in the process of being expanded to include, the conterminous United States (3071 county units). More comprehensive coverage exists for the southern 16 states (1388 county units). The Federal Information Processing Standards (FIPS) state and county codes are used exclusively for county geo-codes.

The temporal aspect of each data set was selected to reflect the most current conditions. In the case of rapidly changing resources, data sets contain observations from a series of years. The reference year for most files lies between 1965 and 1978. Agricultural productivity, bird populations, air quality, and population are examples of files with several years of data. Land use changes occur very rapidly for certain areas, but current data are not available at the county level to define these changes. Climatic data sets include monthly 30-year norms to allow definition of the annual climatic cycle.

The data base is organized into thematic sectors which contain the data sets. For example, the agriculture sector has a data set for crop statistics and another for livestock statistics. Each data set contains a set of variables such as state and county identifiers, reference year, and data items. Data sets are usually created by selecting variables from a single data source. Within the Geoecology Data Base, data sets consist of either one or multiple observations per county unit to minimize storage requirements. With single observations per spatial unit, each computer record contains all the variables for a county. As an example, within the forest resource data set, each county has an observation containing acreage estimates for each of the 12 major forest types in the East, even though not all forest types occur in each county. Data sets with multiple observations per spatial unit include mammal, bird, tree, and vegetation lists. In these cases, there is an observation per county unit for each species or type that occurs within the county, rather than a single observation containing a variable length species list. Data sets with multiple observations are often very large and stored on magnetic tapes rather than on on-line disk units.

The current status of the Geoecology Data Base is summarized in Tables I and II, which indicate types of files, number of variables, and geographic coverage. Table I lists files with one observation per county while Table II includes files with multiple observations per county. Using the average number of observations per county for data sets with multiple observations, there are currently over 1000 variables stored per county unit. For the South there are an additional 500 variables per county. In addition, the Geoecology Data Base includes dictionaries to provide English translations of the codes used in the system, and index files to define regions in which individual county units are located. Indexes exist for climatic divisions, crop-reporting districts, land resource areas, forest inventory units, air quality control regions, water subareas, and others.

### Computer System

The Statistical Analysis System (SAS) (Barr et al. 1978) provides a user-oriented file management system with extensive

Table I
Numbers of variables by geographic coverage for those data sets in the Geoecology Data Base arranged with one observation per county unit.

| Thematic sector | Geographic coverage* South | East | U.S. | Total variables |
|---|---|---|---|---|
| Agriculture | 24 | 73 | | 97 |
| Air quality | 15 | | | 15 |
| Base data and indices | 6 | | 42 | 48 |
| Climate | | 72 | 90 | 162 |
| Energy | 55 | | 155 | 210 |
| Forestry | 49 | | | 49 |
| Land | 21 | | 64 | 85 |
| Population | 25 | | 44 | 69 |
| Terrain | | 46 | | 46 |
| Total variables for data sets with one observation per county unit | 195 | 191 | 395 | 781 |

*South — 16 southeastern states; East — 37 eastern states; U.S. — 48 conterminous states.

Table II
Average number of variables for data sets within the Geoecology Data Base arranged with multiple observations per county unit and covering the conterminous United States.

| Data set | Number of variables per observation | Average observations per county unit | Average variables per county unit |
|---|---|---|---|
| Breeding bird surveys[1] | 8 | 100.0 | 305 |
| Ecoregions | 6 | 2.4 | 8 |
| Endangered species | 6 | 2.2 | 8 |
| Land capability by land use | 28 | 12.3 | 281 |
| Mammal species ranges | 6 | 8.2 | 4 |
| Soil[2] | 8 | 2.0 | 11 |
| Tree species ranges | 6 | 32.5 | 38 |
| Vegetation, potential | 8 | 2.5 | 13 |
| Vegetation, adjusted | 8 | 2.5 | 13 |
| | | | 692 |

[1]Breeding bird surveys are based on 1825 routes within the United States.
[2]Soils data cover only the eastern states.

statistical capabilities (Olson and Strand 1978). Interactive editing and retrieval are available in SAS. SAS also has the flexibility to interface user-written programs to the system for direct access by a SAS program. Each file in the data base is stored as a SAS data set with the directory that contains descriptive information for each variable. The 55 on-line data sets occupy over 10 million bytes of storage.

SYMAP is used to produce contour maps utilizing county centroids as data points and county statistics for data values. Conformant maps consisting of shaded county polygons are produced with the EZMAP program, developed by the Geographic Systems Group at ORNL. All storage and processing are accomplished on an IBM 360/91 computer at ORNL.

## Data Resources

An extensive effort is required to find appropriate data sources and to reformat them to create a data base such as the Geoecology Data Base. This has also been the experience of other agencies requring county-level thematic data to conduct regional studies (Morris and Novak 1979). The Interlaboratory Working Group for Data Exchange (IWGDE) was established in 1976 to increase the sharing of data resources among the eight DOE-funded laboratories (Benkovitz 1979). Based on the Geoecology Data Base experience and IWGDE participation, inventories of available thematic data, problems encountered in large thematic files, and data exchange standards being developed by IWGDE are discussed.

Information centers are able to automatically search key words and abstracts for books and literature on a variety of topics. However, they do not usually contain descriptions of thematic digital files. Such files are usually created and maintained, and not widely documented, by mission-oriented agencies to fulfill their specific needs. Diligent searching and personal contacts are required to find these unique and very useful data sets.

Inventories of spatial data at ORNL have been published periodically since 1975 (Shriner et al. 1978) and a combined inventory of 820 data bases, graphics packages and models from the eight national laboratories was published last year (Shriner and Peck 1978). These and similar inventories provide a means of locating potential sources of thematic data files. However, it must be appreciated that the custodians of most of these files are actively involved in research and development programs and are not oriented to fulfilling requests for data files.

Our experience with extant data has emphasized the need for extensive data checking. Data problems encountered include erroneous data, missing data, wrong county identifiers, and inconsistent county identifiers such as geo-codes for independent cities or other areas, e.g., Yellowstone National Park. The most fruitful ways of determining errors are: using the data, generating thematic maps, searching sorted lists, calculating univariate statistics, and associating two or more independent files. Although users of a data set may eventually find errors, we have found that a preliminary thematic map of a variable will quickly display possible inconsistencies through irregularities in the spatial pattern. In addition, SAS is used to generate ordered listings and simple statistics which often indicate questionable values.

One of the most thorough methods for discovering errors is to combine two independent files by county and check for inconsistencies. Based on ecologic relationships, tree species occurrence can be checked against potential vegetation types within a county; e.g., many tree species do not grow in areas classified as prairie. Another example of cross-file checking would be comparing county acreages as given in most files. Most files containing an estimate of total county area do not agree with the 1970 census values used as our base county acreage. Some of these differences stem from the classification of small and large water bodies within an inventory, or from real changes in county boundaries. However, in other cases it has been exceedingly difficult to resolve differences in the elementary statistic of county size.

The experiences described above have led to the development of data exchange guidelines by IWGDE. Members of IWGDE have published data exchange standards (Benkovitz and Wiley 1977, Brooks et al. 1978) and are extending these conventions to accommodate various geographic based files. These guidelines specify the format for a data definition file which describes the contents of the associated data file. The software to write and read the exchange files has been implemented for IBM, CDC, PDP, and NOVA computers as a prerequisite of efficient data exchange. The adoption of the current standard as a formal ANSI Standard is under consideration by the American National Standards Committee, X3L5 (Information Interchange Data Descriptive File ANSI X3L5/78-62).

To minimize editing and reformatting efforts, the IWGDE has drafted a set of conventions for the content of thematic files. Conventions include exchange standard format; five digit FIPS geo-codes for county units; definition of independent city/county aggregations; metric units of measure; missing value codes; and documentation standards (IWGDE 1979). Files converted to conform to these conventions would be designated as reference data bases, and these would be created by individuals familiar with the subject content and actively using the file, so that errors and inconsistencies would be minimized. Table III lists currently available data bases which were considered by IWGDE to be essential to regional studies. IWGDE is not itself creating this set of reference data bases, but it has proposed the conventions which would result in a set of compatible, useful, thematic data bases for the nation.

## *APPLICATIONS OF THE GEOECOLOGY DATA BASE*

### Air Pollution Impacts

One of the activities of the Environmental Sciences Division is to predict and evaluate environmental impacts from energy technologies. As part of this role, ESD studied possible future environmental impacts in the South from the expanded use of coal. One of the primary impacts of burning more coal appears to be increased levels of sulfur dioxide ($SO_2$), which causes injury to certain agricultural crops and forest (timber) resources.

To estimate the potential impact of increased levels of $SO_2$ on vegetation, we employed the use of maps generated from long-range trajectory models based on coal-use scenarios for 1985 (Davis 1978). Estimates of $SO_2$ levels were transcribed from a rectangular grid to county centroid coordinates and mapped with SYMAP (Figure 2a). Since it was almost impossible to ascertain what effect ambient $SO_2$ levels would have on macro-scale systems, the map of predicted 1985 levels was overlaid with maps of the vegetation most likely to be impacted. This was done for soybean cropland (Figure 2b) and softwood forests (predominantly pine species), because they are known to be sensitive to $SO_2$ pollution. The darkest regions on the resulting maps indicate the coincident areas with the highest crop productivity (Figure 2c) or highest softwood productivity (Figure 2d) and the highest levels of atmospheric $SO_2$. While the role of emissions from coal-fired electric plants in the magnitude and occurrence of these $SO_2$ levels may be debated, there is a high probability that the effluents from steam electric plants will interact with those from other industrial and urban sources. Thus, while the $SO_2$ levels depicted here may not cause acute injury to vegetation, they may, when combined with ambient $SO_2$ levels, cause severe chronic injury. Furthermore, levels represent annual average for county cells, so that there may still be periodic episodes resulting in acute injuries to the vegetation. The darkest areas on the computer-generated maps indicate areas of justifiable concern. For example, if the counties which are darkest on the $SO_2$-soybean map (Figure 2c) were to suffer a 10% drop in yield, the result would be a loss of 17 million bushels, or over $100 million per year. Although the results are very general at this point, they indicate a substantial potential impact of $SO_2$ on pine forests and soybean yields. This coupling of the Geoecology Data Base to the synagraphic mapping systems demon-

Table III
Extant Spatial Data-Bases Proposed by the DOE Interlaboratory Working Group for Data Exchange as Possible Reference Data Bases

| | Year | Geo unit | Original source |
|---|---|---|---|
| 1. Population (by Age, Sex, Race) | 1950 | Counties | Census |
| | 1960 | Counties | Census |
| | 1970 | Counties | Census |
| | 1975 | Counties | Census |
| 2. County and City Data Book | 1972 | County, Cities | Census |
| | 1977 | County, Cities | Census |
| 3. MED-X | 1970 | County, Cities | Census |
| 4. Mortality Rates (53 causes) | 1968-1972 | Counties | NCHS |
| 5. Air Quality Control Regions | 1975 | County | EPA |
| 6. Point Source Emissions | 1974-1976 | Aggr., Points | EPA |
| 7. Ambient Air Quality | 1974-1976 | Points, AQCR's | EPA |
| 8. PARAP (Population at Risk to Air Pollution) | — | Counties | |
| 9. Weather Station Locations | — | Points | NOAA |
| 10. Weather Surface Observations | 1880's-1978 | Points | NOAA |
| 11. Monthly Temperature and Precipitation for State Climatic Divisions | 1931-1975 | County Aggregates | NOAA |
| 12. Wind Roses | — | Points | ? |
| 13. Watershed Boundaries | — | Polygons | USGS |
| 14. Water Quantity | | Points | USGS, EPA |
| 15. Water Quality (NASQAN) | | Basins | USGS, EPA |
| 16. Land Use from Conservation Needs Inventory | 1967 | Counties | USDA |
| 17. Agriculture Production | 1974 | Counties | USDA |
| 18. Endangered Species | 1978 | Counties | USF&WS |
| 19. BNL Energetics Atlas | 1972-1974 | Counties | BNL |
| 20. Coal Reserves | — | Counties | DOE |
| 21. County Outlines | 1970 | Counties | |
| 22. Census Tract Boundaries | 1970 | Tracts | |
| 23. World Data Bank II | 1977 | | |

strates not only the applicability of the data base, but also the versatility of the combined systems for regional analysis.

### Natural Vegetation and Land Use

The need to incorporate information on land use and vegetation into regional studies prompted the use of extant data to determine the amounts and types of natural vegetation existing within the United States. The preservation of natural vegetation is an integral part of the U. S. Forest Service's Roadless Area Review and Evaluation program (RARE-II) (USFS 1978). This program allocates roadless areas to wilderness as part of the National Wilderness Preservation System or to other land uses based on many criteria. In an ecological analysis of RARE-II tracts (Klopatek et al. 1979a), we developed estimates of natural vegetation types adjusted for current land use. The amount of a particular vegetation type occurring on a RARE-II tract was compared to the adjusted estimate of the amount of that vegetation type occurring in the United States. This measure of vegetation uniqueness was one of several factors used to identify ecologically similar tracts which could be used to resolve land use conflicts while preserving all natural vegetation types.

The determination of the loss of natural vegetation was accomplished in four steps. The first involved acquiring county information on the size of the area which could be occupied by each natural vegetation type. Vegetation data were digitized from Küchler's (1964) map of potential natural vegetation of the United States. The area of each vegetation type in each county in the United States was measured by using either planimeter or a dot grid method. These data were then stored in the Geo-Ecology Data-Base.

The second step was to obtain current estimates of land use practices for county units. The 1967 Conservation Needs Inventory (CNI) (USDA 1971) land use data, stored in the data base, were aggregated to give land use estimates for agriculture, pasture, range, forest, other (nonfederal rural land that is not classified as cropland, pastureland, rangeland, or forest), urban build-up, water, and federal land. Urban, water, and federal subdivisions of the noninventory acreage were obtained from Iowa State University Statistical Laboratory. The acreages were normalized to the 1970 census county acreage.

In the third step, a set of probabilities were assigned to vege-

*Cont'd. page 102*

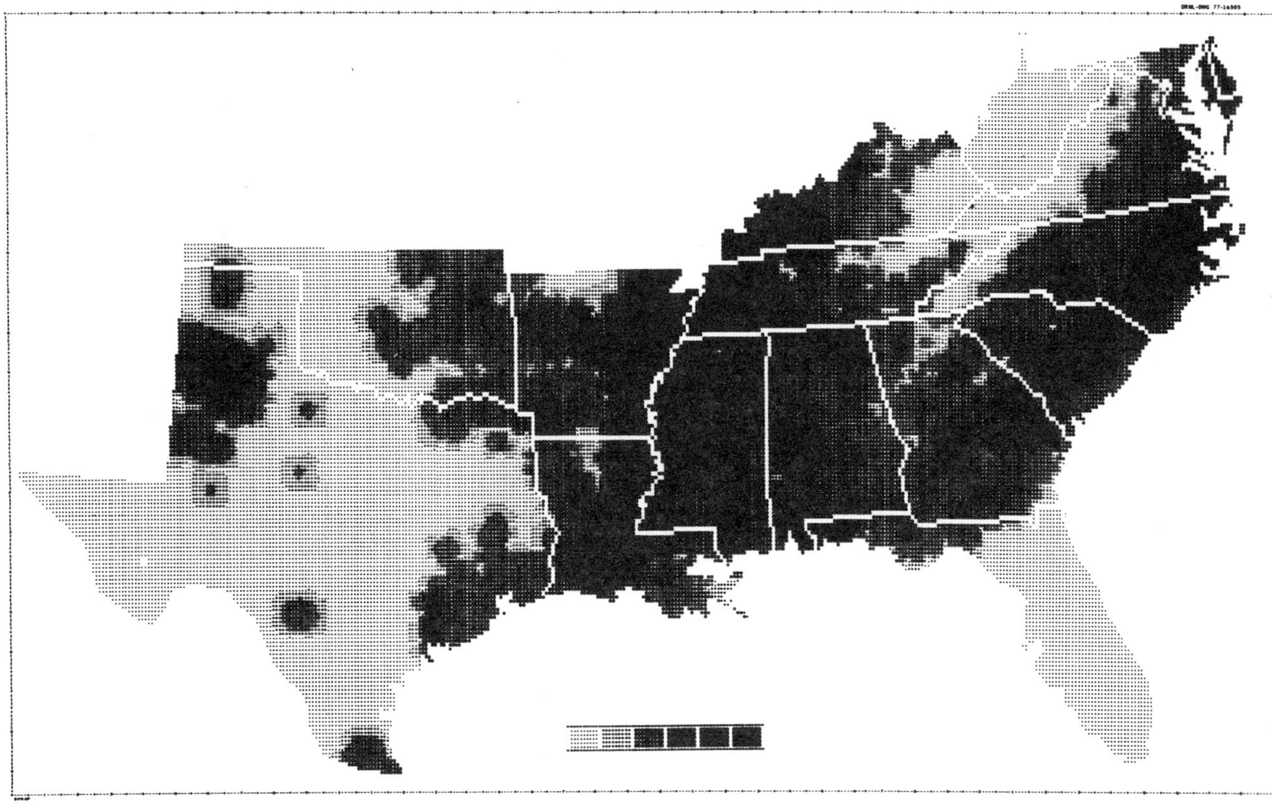

Figure 2a.

Figure 2b.

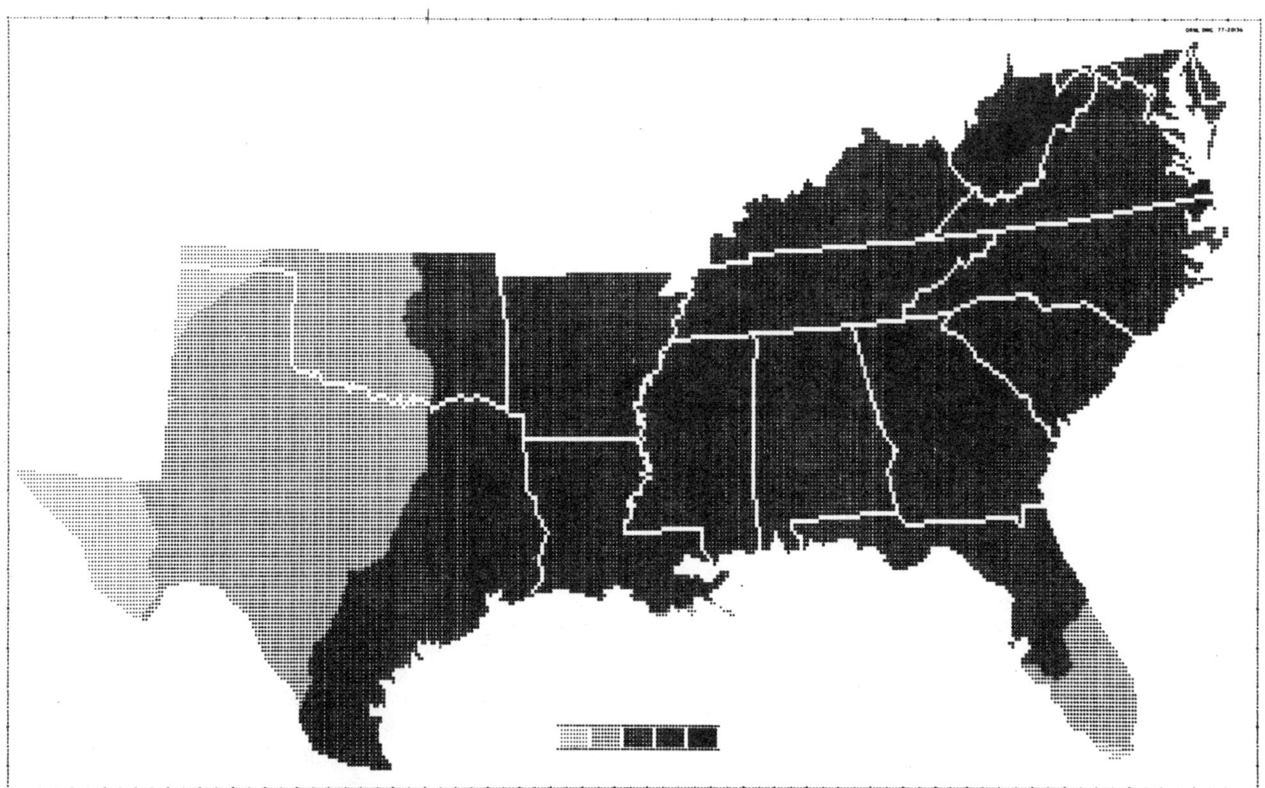

*Figure 2. Evaluation of Potential Impact of Predicted 1985 $SO_2$ Concentrations on Soybeans and Softwood Forests in the Southern United States (Davis 1978). Darkest Areas on Maps Represent Highest Concentrations (Figure 2a), Yields (Figure 2b), or Impact (Figures 2c and 2d).*

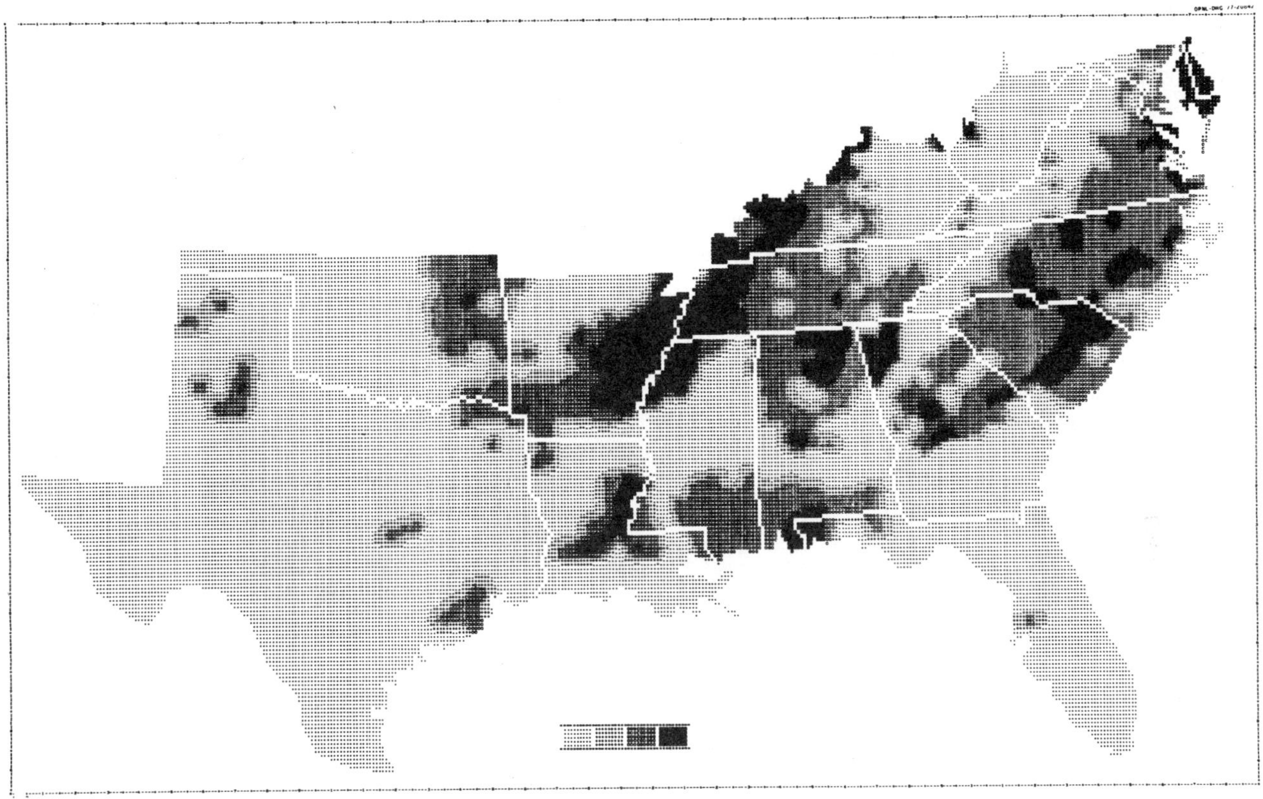

*Figure 2c.*

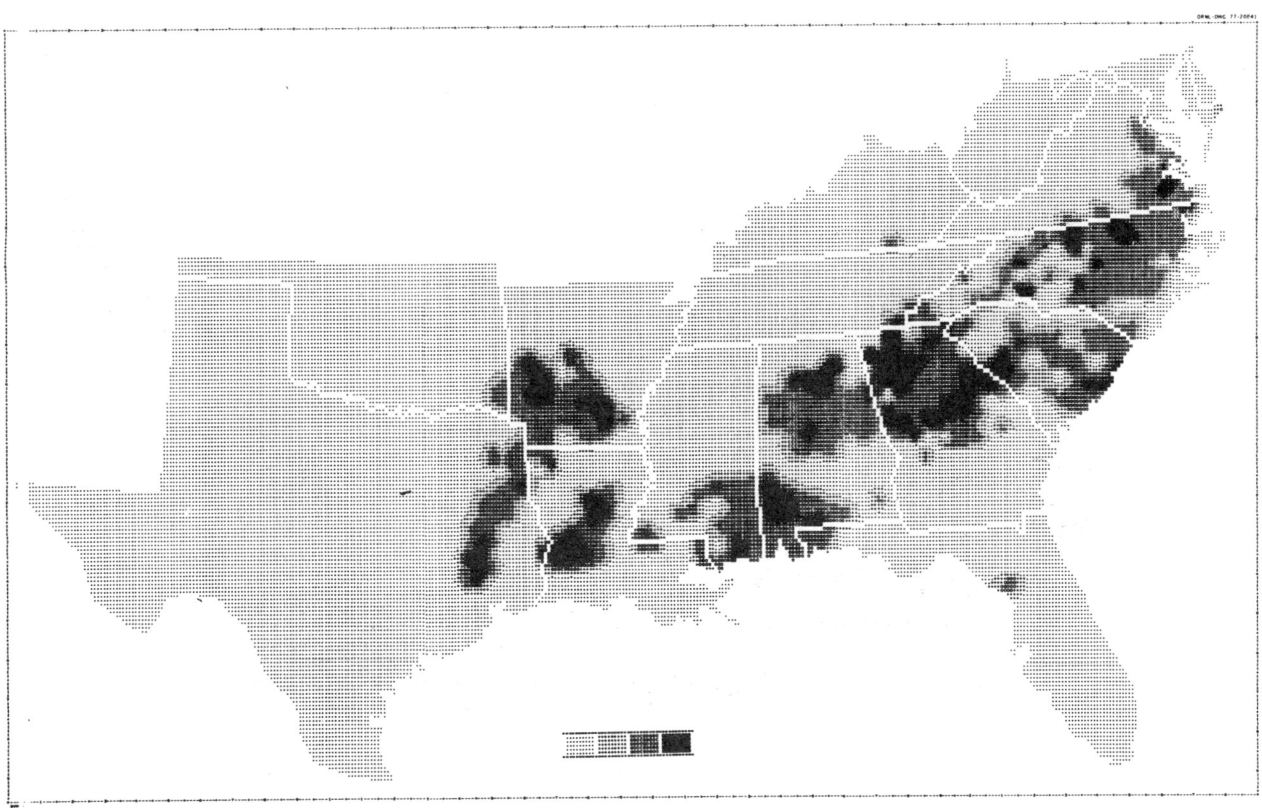

*Figure 2d.*

tation types which defined their probability of conversion to other land use types. The probabilities of being converted to agricultural land, pasture land or water areas (reservoirs) were assigned to the respective vegetation types based on a number of sources (Klopatek et al. 1979b). For all large western urban areas (>10,000 hectares) the county urban areas were keyed, through visual interpretation, to the potential vegetation that they replaced. For all counties containing urban areas in the vegetationally more homogeneous, smaller eastern counties, the total urban land was subtracted proportionally from the potential vegetation types occurring in the county. The same was done for the CNI "other land" category.

Finally, a FORTRAN computer algorithm was developed to subtract the CNI county land use data from the appropriate potential natural vegetation types occurring within the county. Figure 3 graphically depicts the percentage of potential natural vegetation still existing in the conterminous United States today. The results give a fairly accurate and indeed startling portrayal of the area occupied by natural vegetation. The spatial pattern of remaining natural vegetation areas reinforces the suggestion that preservation action is urgently required to protect many vegetation types. The entire results of this study are discussed in a paper by Klopatek and others (1979b). From a pragmatic viewpoint this example represents a melding of extant and empirical data within the data base to produce results which can be used to analyze and assess regional environmental problems.

### Wilderness Area Evaluation

As mentioned previously, ESD was involved in ecologically rating the 2686 U. S. Forest Service's roadless tracts which totalled nearly 62 million acres in the national forests. The Department of Energy, which was concerned with both the energy resource potential and the environmental values of the RARE-II sites, funded this project to evaluate these values for the tracts, which ranged in size from one to several hundred thousand acres and occurred from Puerto Rico to Alaska. Evaluation of these tracts required the development of a sophisticated methodology as well as extensive data bases.

Expression of the ecologial value of a wilderness area involved identifying important natural resources located within the area; determining the magnitude of those resources; and computing the area quality value. The area quality value combines the magnitude of a particular resource with the importance of that resource. This procedure results in a value which can be compared to similarly derived values for other sites.

Based on a number of considerations, four major parameters were chosen to construct a quantitative ecological index of a tract of land located anywhere in the United States (Klopatek et al. 1979a). These were: (1) vegetation: the vegetation communities according to Küchler's (1964) system, adjusted for current land uses as discussed in the previous section; (2) avian communities: the species and abundance of birds; (3) mammal communities: the species of mammals present; and (4) endangered or threatened species: the occurrence of such

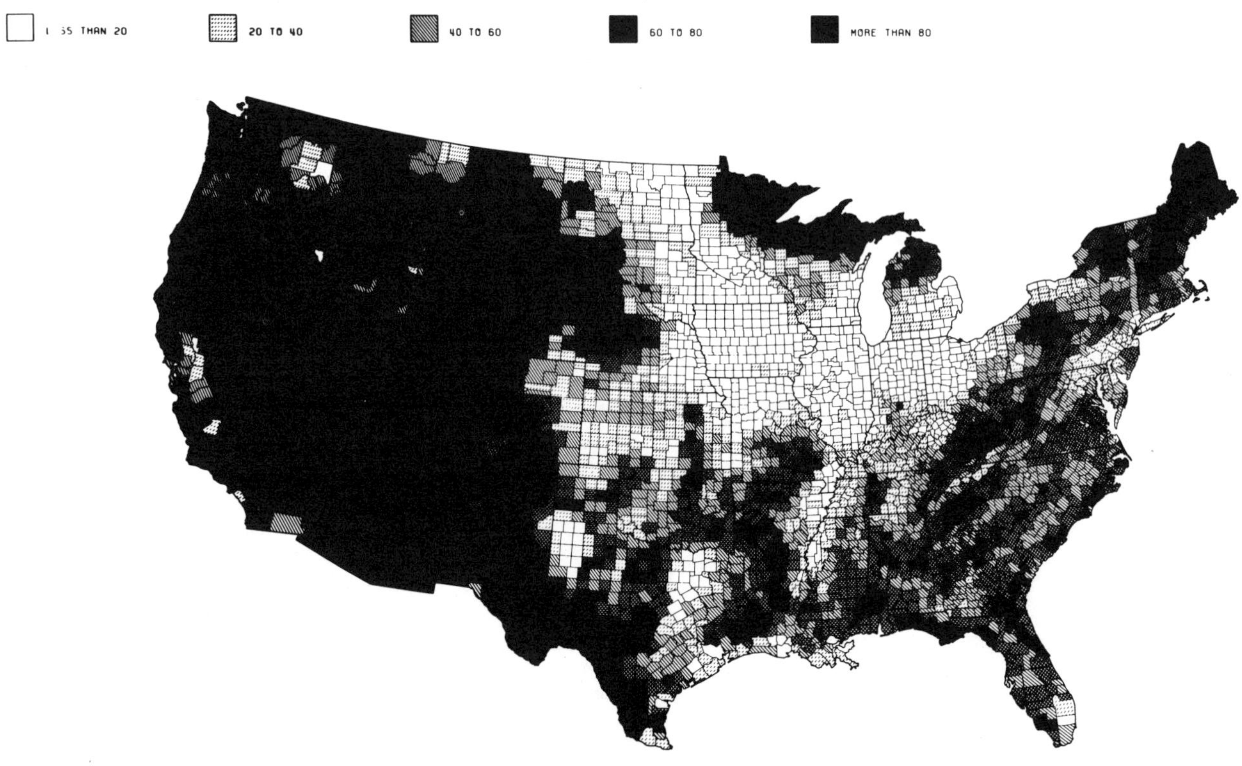


*Figure 3. Remaining Potential Natural Vegetation for Counties in the United States (Klopatek et al. 1979b).*

species and whether or not the area provides critical habitat for the species.

Each wilderness tract was evaluated within a hierarchical framework in order to provide the capacity to search for tracts of similar value when wilderness/development conflicts arose. All parameters were evaluated at national, regional, eco-region sectional, and individual tract levels. The 54 eco-regions of the United States (Bailey 1976) represent broad areas considered to be ecologically homogeneous and were used to define ecological strata within which alternative tracts could be selected. The combining of values for various levels provides a measure of how a particular tract's resources relate to the whole county. Tracts containing a nationally or regionally rare resource would be given a high rank for that attribute. Mathematically, the ecological index reflects a filtering of information from one level to another. Thus, by combining values for all the parameters at all levels, an overall ecological index can be calculated for each roadless tract.

An application of the methodology is shown for eco-region M3112, the Douglas fir region of the United States (Figure 4). This figure displays the overall ecological indices, which are composites of the vegetation, mammal, avian, and endangered species rankings. In Figure 4 the size of the circles is proportional to the size of the tracts. The most dense circles indicate that the tract ranked in the upper 20% of all tracts; the open circles indicate a ranking in the lower 20%. As can be seen, the upper level ratings of the tracts are area and state independent. These ecological ratings were then compared with the energy rating of these tracts (this area contains the Rocky Mountain Overthrust Belt, a region claimed by industry to have the potential for greater oil and gas reserves than the Alaskan North Slope). Potential conflicts in tracts which received both high energy and high ecological ratings were noted for alternative or mitigation strategies. The methodology developed for the ecological analysis and evaluation of the RARE-II tracts can thus be viewed as a trade-off analysis to help resolve environmental and energy resource conflicts. This application demonstrates the flexibility of the Geoecology Data Base in combining statistical analysis and computer mapping. It shows the relative comparison of tracts of land which are either subcounty in size or multicounty in area.

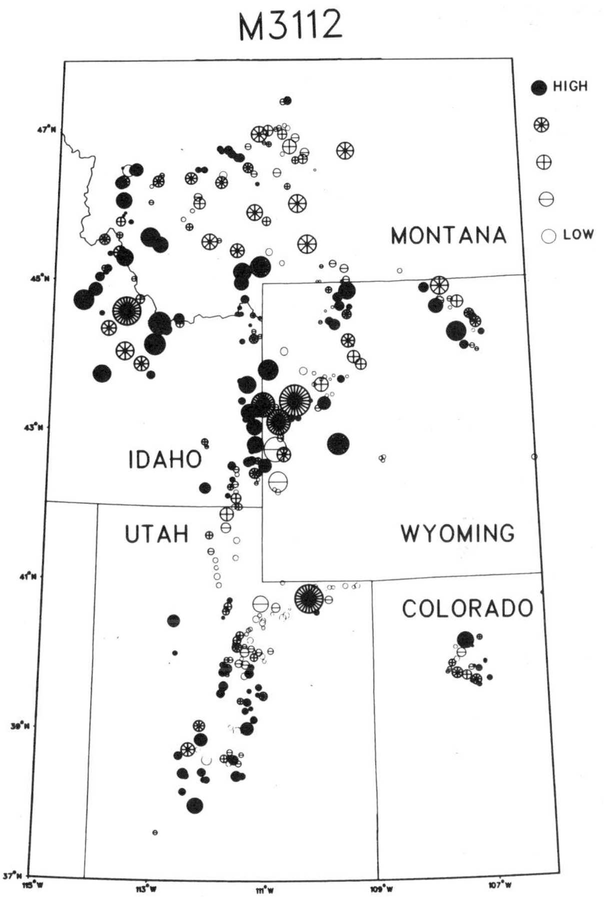


*Figure 4. Overall Ecological Indices of the RARE-II Tracts within the Rocky Mountain Overthrust Belt (RMOB) and the Douglas Fir Eco-Region M3112. Index Symbols Are Displayed According to Ranked Percentiles (0-20, 21-40, 41-60, 61-80, 81-100) (Klopatek et al. 1979a).*

## DISCUSSION

In October 1975, a week-long workshop on energy-related regional analysis was held in Oak Ridge, Tennessee, with participants from most of the DOE-funded national laboratories. A task force on regional environmental data bases within the workshop drafted a statement stressing the need for interagency cooperation and coordination in developing a national environmental "data base." The goals were to minimize costly duplication of inventory and data management efforts, provide needed information for rational resource development, and promote optimal resource inventory design. The task force recommended a "distributed data base" concept with the extensive data files continuing to reside in the responsible agencies. Creating a huge centralized data center was not considered feasible or desirable. Those agencies involved in environmental analysis and assessment, such as DOE, EPA, and CEQ, could then assemble and synthesize a data base of selected variables from the distributed data files. The resulting data base should have quick response capability, spatial and temporal standardization, and high exportability.

The Geoecology Data Base represents a county-level file created by selecting variables from distributed data bases, with extensive editing required to standardize the data base. We demonstrated by example that county-level data bases can be flexible enough to meet the needs of various regional studies. Furthermore, we described the conventions for data exchange and standardized reference data bases that are being implemented by the DOE Interlaboratory Working Group for Data Exchange. These guidelines are available for use by individuals creating and willing to share thematic data files. With additional review, these standards can be adopted by the community of agencies involved in compiling and applying these data bases. With these standards, updating and expanding the Geoecology Data Base or similar data bases with new data will become a trivial task. In light of increasing needs for thematic spatial data, it is appropriate that conferences, such as the Harvard Computer Graphics Week, provide a mechanism for increasing data sharing activities. Descriptions of available data bases need to be shared and initial steps need to be taken in establishing standards for a coordinated national environmental "data base."

## *BIBLIOGRAPHY*


Bailey, R. G. 1976. *Ecoregions of the United States.* U. S. Forest Service, Ogden, UT.

Barr, A. J., J. H. Goodnight, J. P. Sall, and J. T. Helwig. 1976. *A User's Guide to SAS 76.* SAS Institute, Inc., Sparks Press, Raleigh, NC.

Benkovitz, C. M. (ed.). 1979. "DOE Interlaboratory Working Group for Data Exchange Progress Report — Oct. 1978." Brookhaven National Laboratory, Upton, NY (in press).

Benkovitz, C. M., and R. A. Wiley. 1979. *Using System 2000 with Proposed ANSI Standards for Data Exchange.* LA-UR-77-2739. Los Alamos Scientific Laboratory, Los Alamos, NM.

Brooks, A. A., F. D. Hammerling, and B. N. McNeeley. 1978. *User's Guide for an IBM PL/1 Implementation of Level 1 of the Proposed American National Standard Specifications for an Information Interchange Data Descriptive File.* ORNL/CSD/TM-74. Oak Ridge National Laboratory, Oak Ridge, TN.

Burgess, R. L., and L. H. Kern (eds.). 1973. *Progress Report 1971-72, Eastern Deciduous Forest Biome, US-IBP.* EDFB-IBP/73-5. Oak Ridge National Laboratory, Oak Ridge, TN.

Council on Environmental Quality (CEQ). 1979. "The County-Level, Integrated Data Extraction (CLIDE) Data-Base User's Guide." President's Council on Environmental Quality, Washington, DC (draft report).

Davis, R. M. (project coordinator). 1978. *A Preliminary Regional Assessment of Coal Utilization in the South.* Volumes I and II. ORNL/TM-6122. Oak Ridge National Laboratory, Oak Ridge, TN.

Klopatek, J. M.; J. T. Kitchings; K. D. Kumar; and R. J. Olson. 1979a. "An Ecological Analysis of the U. S. Forest Service's RARE-II Sites." ORNL/TM-6813. Oak Ridge National Laboratory, Oak Ridge, TN (in press).

Klopatek. J. M.; R. J. Olson; C. J. Emerson; and J. L. Joness. 1979b. "Land Use Conflicts with Natural Vegetation in the United States." Environmental Conservation (in press).

Küchler, A. W. 1964. *Potential Natural Vegetation of the Conterminous United States.* American Geographical Society Special Publication. No. 36. New York.

Morris, S. C., and K. M. Novak (eds.). 1979. "Use of County Level Data in Health, Energy, Demographic, Environmental and Economic Analysis." *Proceedings of a Computer-Based Conference,* Mar. 1977. Brookhaven National Laboratory, Upton, NY (in press).

Olson, R. J., and F. G. Goff. 1976. "Development and Applications of a Regional Environmental Data Base for Southeastern United States." pp. 289-293. Presented at: Fifth International CODATA Conference, Jun. 27 — Jul. 2, 1976, Boulder, CO. Published in: *The Proceedings of the Fifth Biennial International CODATA Conference,* edited by B. Dreyfus.

Olson, R. J., and R. H. Strand. 1978. "Management of Diverse Environmental Data with SAS." pp. 200-206. In *Proceedings, Third Annual Conference, SAS Users Group, International.* SAS Institute, Inc., Raleigh, NC.

Shriner, C. R. (ed.); A. A. Brooks; R. M. Davis; F. A. Heddleson; P. E. Johnson; C. A. Little; A. S. Loebl; R. J. Olson; L. J. Peck; B. C. Talmi; J. A. Watts; and D. L. Wilson. 1978. *Spatial Data on Energy, Environmental, and Socioeconomic Themes at Oak Ridge National Laboratory; 1977 Inventory.* ORNL-5395. Oak Ridge National Laboratory, Oak Ridge, TN.

Shriner, C. R., and L. J. Peck. 1978. *Inventory of Data-Bases, Graphics, Packages, and Models in Department of Energy Laboratories.* ORNL/EIS-144. Oak Ridge National Laboratory, Oak Ridge, TN.

U. S. Department of Agriculture (USDA). 1971. "National Inventory of Soil and Water Conservation Needs, 1967." Statistical Bulletin No. 461.

U. S. Forest Service (USFS). 1978. RARE-II. *Draft Environmental Statement. Roadless Area Review and Evaluation.* USDA Forest Service. Jun. 1978.

# Use of LANDSAT Data to Expedite Soil Survey Operations

*by John B. Peterson*

## PROBLEM DESCRIPTION

The mapping of soil, soil erosion, and land use has developed as one of the most effective ways of identifying, delineating, and portraying land resources. Although aerial photography from balloons was used in the mid-nineteenth century, especially in the Civil War, the feasibilities were not seriously explored until World War I. After that war soil scientists quickly experimented in the use of aerial photography as a basis for mapping soils.[4 6 7 8 12 14] In 1929, T. M. Bushnell[3] reported the first aerial photographing of an entire county for the cooperative soil survey of the US Department of Agriculture and the state experiment stations.

Early in the 1960's certain discrete wavelength bands and combinations of these bands were found to be more discriminating in characterizing soil differences than were the usual broad spectrum photographic images. Condit,[5] using a laboratory spectrophotometer to secure multiband data, found the spectral reflectance of a wide range of soils could be predicted accurately from measurements using five wavelength bands between .32 and 1.0 micrometers. Montgomery and Baumgardner[13] found multispectral measurements with a field spectroradiometer to be largely correlated with cation exchange capacity and silt content. Kristof[10] successfully used data secured by the optical mechanical scanner on an aircraft platform to differentiate levels of soil organic matter in a field of freshly tilled soil.[2] Surface soil areas of different organic matter levels over a 40 km flight line in central Indiana were delineated and successfully mapped with multispectral data and automatic data processing.[11]

Results from the rapidly growing volume of research represented in the foregoing examples led to anticipation of possible benefits for soil survey operations. Consequently, early studies at Purdue Laboratory for Applications of Remote Sensing (LARS) to expedite the delineation of soil associations were tried, but with intermediate success. For example, when LANDSAT imagery was compared to the soil association map of Tippecanoe County, Indiana, after processing both by photographic techniques and by computer analysis, only four groupings of 13 soil associations were separated. This can be explained by the fact that soil associations are not necessarily groups of soil series with similar morphological properties.

The growing costs of mapping soils resulting from inflation prompted continued efforts to determine the usefulness of remote sensing techniques. In 1978 the Soil Conservation Service (SCS) of the USDA mapped 50 million acres using traditional methods based on walking the land to sketch soil type boundaries on an aerial photograph base. The cost of implementing an operation with traditional methods is, of course, quite high.

## METHODS USED AND RESOURCES REQUIRED

The methods used were primarily based on analysis of LANDSAT multispectral scanner (MSS) data covering the study areas by a computer-implemented analysis package, LARSYS.[17] Initially a clustering algorithm was used to divide the MSS data into groups of data points of similar reflectance values. This was accomplished by calculating the Euclidean distances between each point and the initial cluster class center and assigning each sample to the class with minimum distance. Initial cluster centers were selected by considering the mean and standard deviation of all the data points and evenly dividing this distribution into a specified number of segments. After the initial cluster classes were formed, new cluster class centers were developed by considering the mean vector of each initial cluster. This process continued until the cluster class centers did not change from one iteration to the next.[1]

Since the purpose of our study was to seek ways of using remotely sensed data to expedite the soil survey, computer-implemented data analysis was only a part of the approach. A team of specialists gradually took shape who could communicate freely with each other, could articulate the necessary knowledge in their individual specialties in a way their colleagues could understand and who could rapidly come to understand each other's expertise well enough so that the group effort could capitalize on the knowledge within the group. This meant working not only with an interdisciplinary team within LARS but also with personnel of the SCS who were actively engaged in mapping in the field. We were fortunate in receiving complete cooperation from SCS soil scientists who were in the process of mapping the soils in neighboring counties.

## RESOLUTION OF PROBLEMS ENCOUNTERED: DOCUMENTATION, EXAMPLES AND RESULTS

The evaluation of the present effective expedition of soil survey operations through techniques developed at LARS in cooperation with the SCS developed step by step. Encouragement to look at slight differences in soil color resulted from an attempt by a soil scientist and a data analyst to explore the possibility of delineating the slight color differences between udalfs (timber soils) and udolls (prairie soils), which lie on opposite sides of the classical boundary between the northern edge of the "prairie point" and the originally timbered lands of west central Indiana (Figure 1).

Field identification of zonal "udolls" and zonal "udalfs" were made and surface soil colors read against Munsell color charts[15] were recorded. MSS data in the 0.60 to 0.70 and 0.70 to 0.80 micrometer ranges were acquired over sample area 210 by the University of Michigan single aperture twelve-channel MSS aboard the Michigan C47 aircraft during the Corn Blight Watch Experiment. The data were subjected to a clustering techique. The resulting demonstrated capability to separate quantitatively slight differences in surface soil colors representative of these two soil suborders were very encouraging (Figures 2 and 3). Field observations established that the udalfs and udolls when moist differed in Munsell color readings only slightly. Thirty percent of the udoll samples were read as 4/2 in the Munsell chart and 54 percent of the udalfs as 3/2, only one unit apart in value and of identical chroma. Other readings were close to these. The results were of special interest in view of the very slight differences in surface soil colors observed by the human eye between the zonal udoll and udalf soils.

The second encouraging event was the discovery of a three to five mile wide meandering stretch of dark soil running from south of Kokomo to the vicinity of Lafayette, Indiana, and the results of our analysis of this phenomenon.[16] This landform was found to be the surface of a buried tributary of the Teays preglacial river system. Apparently the dark soils resulted from lagooning of the last filling of the valley during late Pleistocene times, resulting in relatively flat, fine-textured soils, which were poorly drained and marshy enough to be anaerobic, with the resulting accumulation of humus characteristic of such soils. These soils were obviously much darker than the rolling, originally timbered soils of the surrounding till plains (Figures 4, 5, 6, 7).

In setting up five classifications based on LANDSAT-1 data

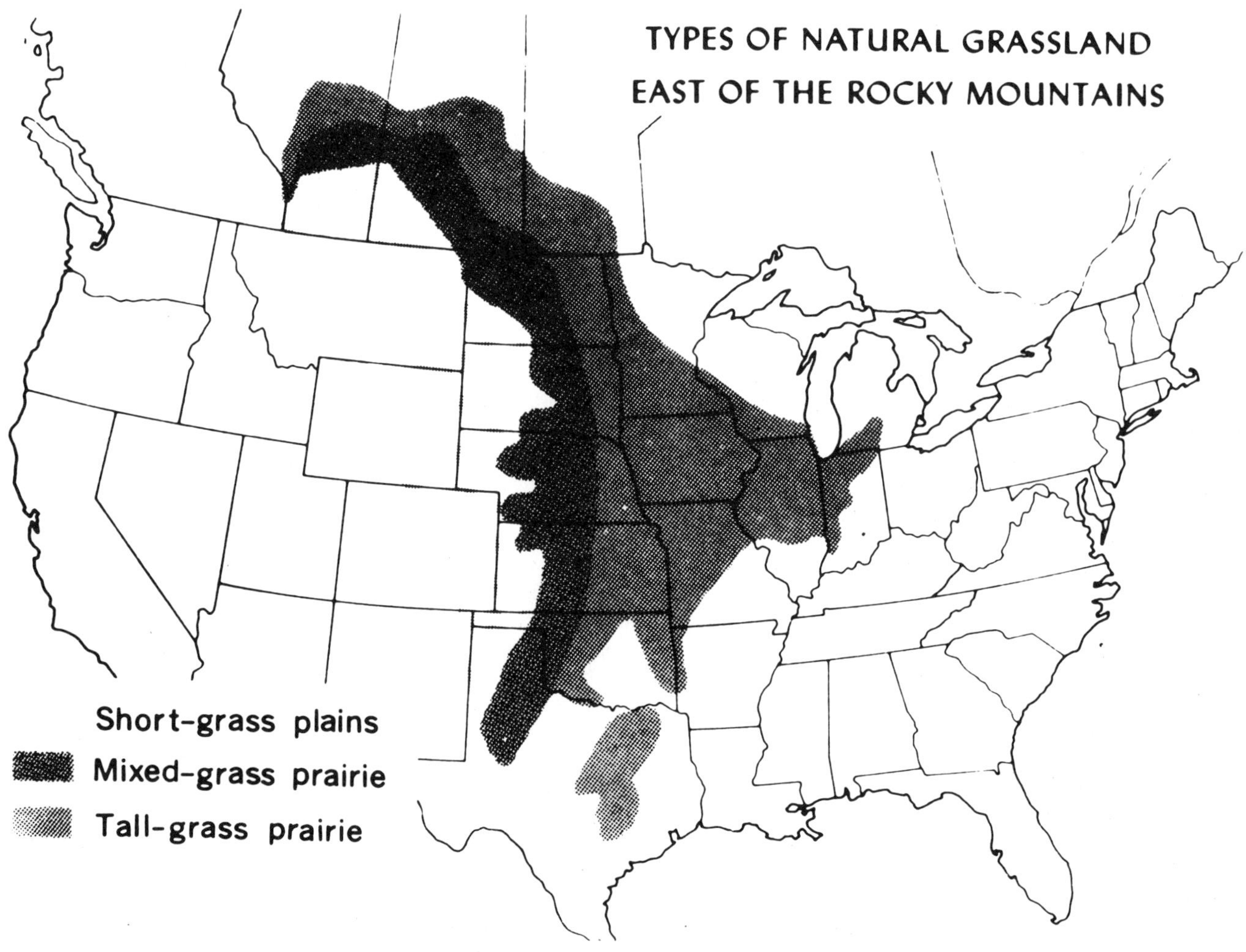


*Figure 1. Prairie Point.*

From *Climate,* by Glen Trewartha; Copyright 1968 by McGraw-Hill Book Company

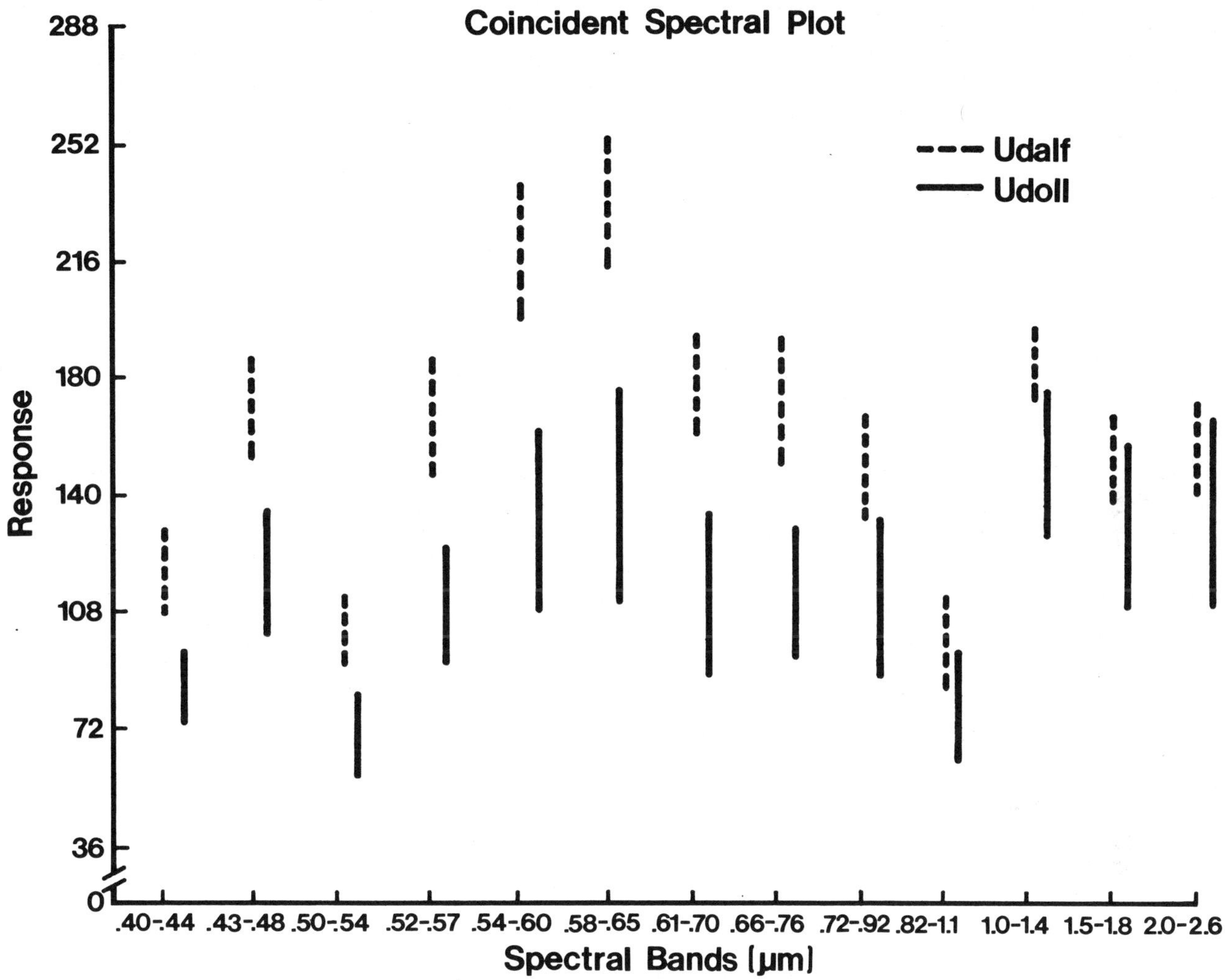


*Figure 2. Relative Reflectance of Udoll and Udalf Soils.*

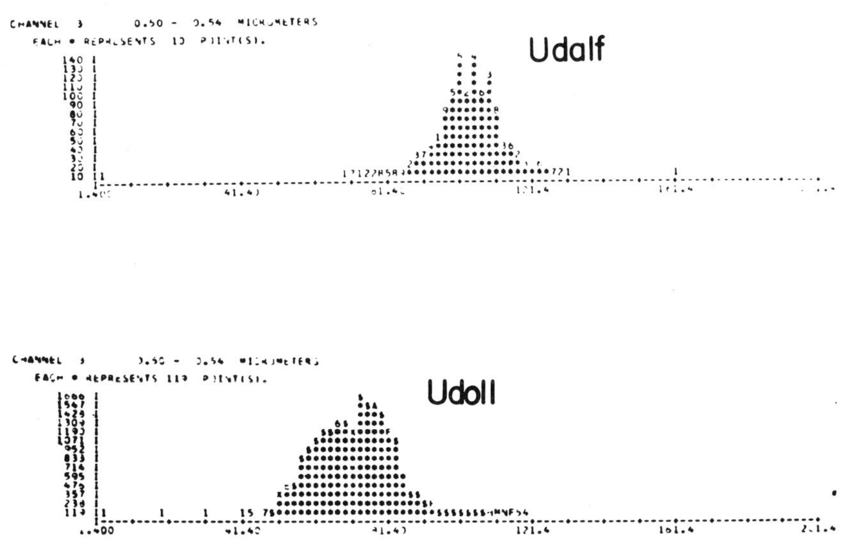


*Figure 3. Histograms of Relative Data Values for Reflectance at 0.50 to 0.54*

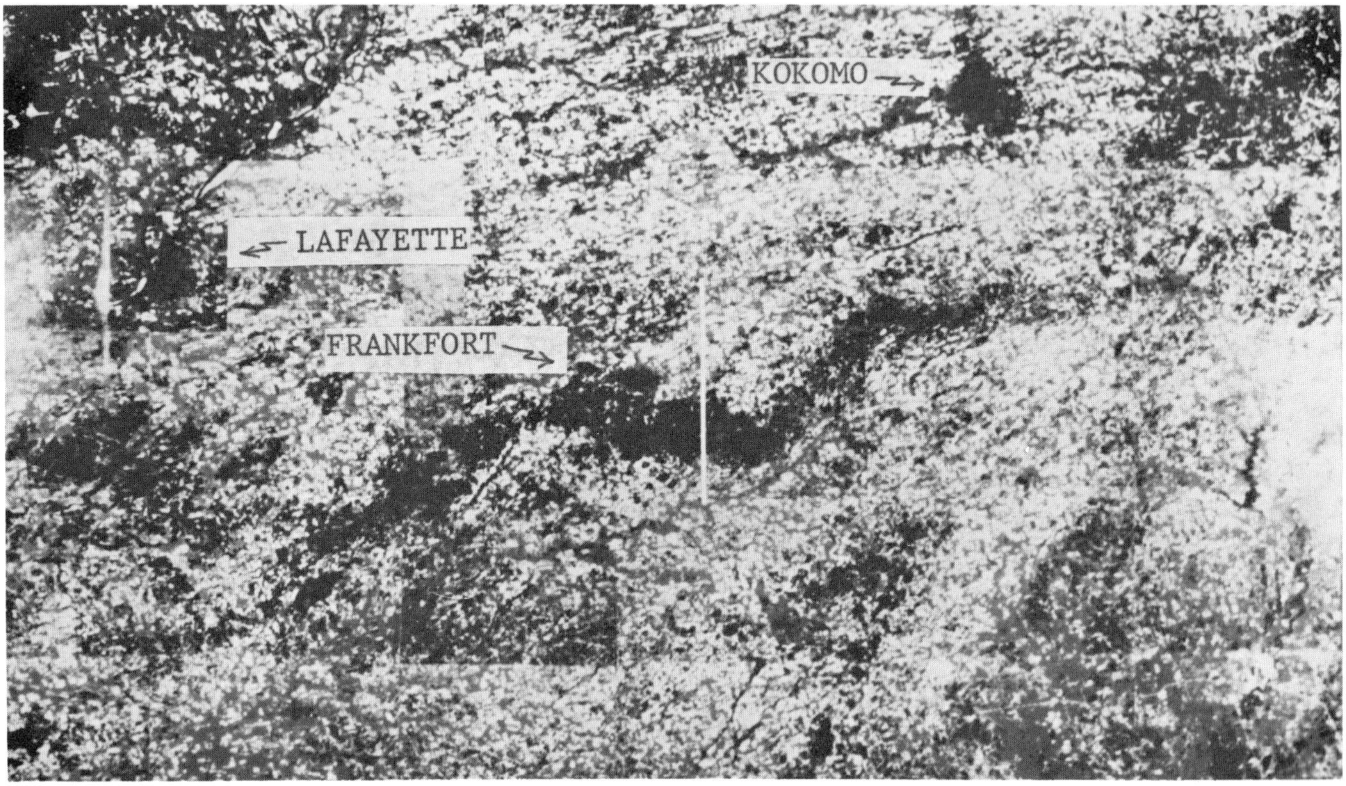


Figure 4. Surface of Buried Preglacial Valley.

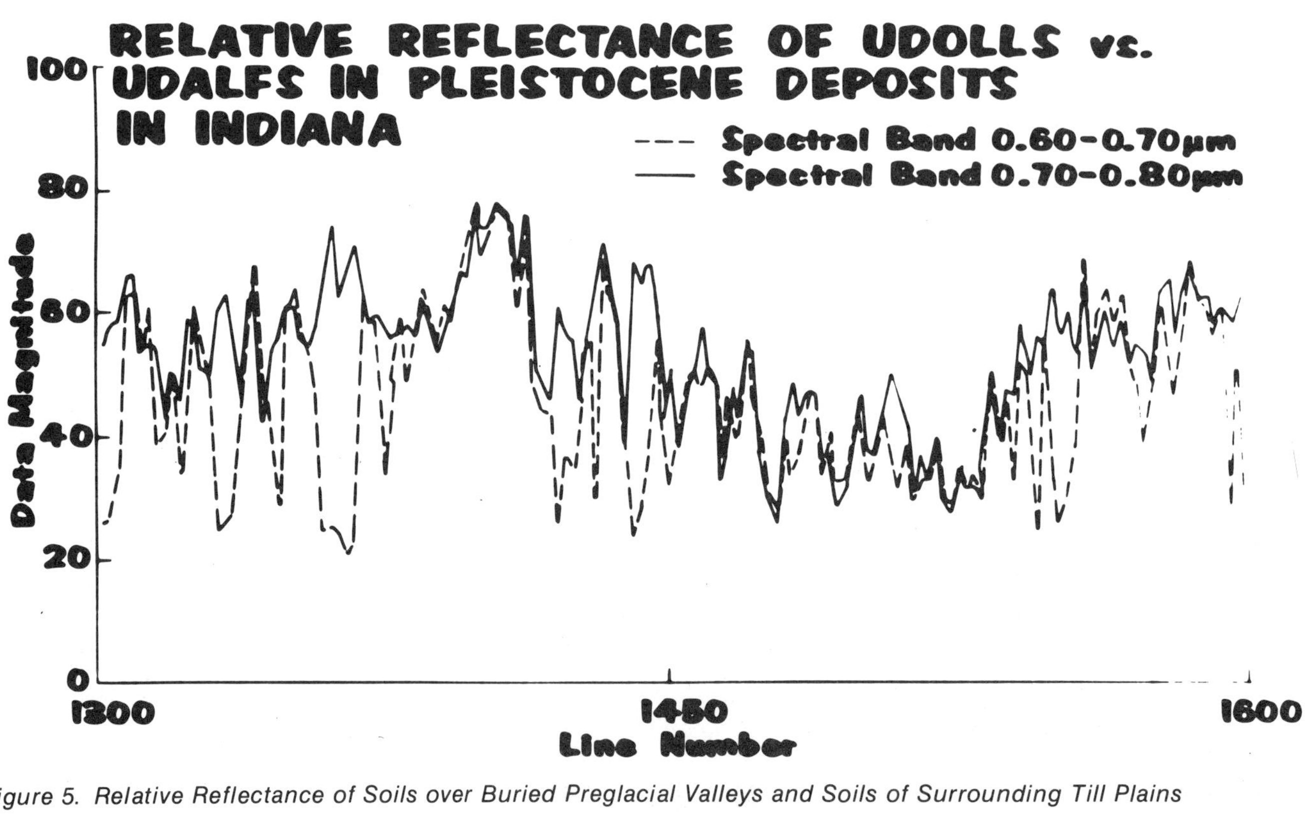


Figure 5. Relative Reflectance of Soils over Buried Preglacial Valleys and Soils of Surrounding Till Plains

*Figure 6. Typical Udoll Landscape of Buried Preglacial Valley.*

*Figure 7. Typical Udalf Landscape on Originally Timbered Till Plains.*

to help us differentiate the soils on the valley fill from those of the adjacent till plains, it became obvious that the three classes assigned to the dark soils were closely delineating important soil series differences. Consequently, a plan was developed for a cooperative effort by LARS and the SCS to explore the utility of LANDSAT data as an aid to the soil survey in progress at that time (1975) in Clinton County, Indiana. A LARS team of two data anlysts and three soil scientists was assigned to work with the SCS soil survey party in Clinton County to determine the possibility of fitting spectral data classes to important soil groupings closely enough for such classifications to be useful to the soil scientist walking the land.

The results were impressive. Early in the study the SCS field men began advising us to look for discrepancies in their maps rather than for errors in the spectral data. They were impressed with the geographic accuracy and the consistency of the spectral data. It was estimated that the delineations in spectral data were so indicative of important soil profile changes that by having at hand a registered, rectified, geometrically correct spectral map the surveyor could complete his soil map in approximately half the usual time.

Furthermore, the spectral map was making possible identification of significant characteristics of the landscape down to a detail of approximately half a hectare. In addition, the quantitative aspects of the spectral data provided an invaluable new dimension to land survey, namely the capability of securing tabular data rapidly in various forms regarding the areas involved. Of special value was the capacity to estimate much more accurately than ever before the percentage of land occupied by an inclusion within a soil type boundary.[9] The value of a spectral map in delineating soil type boundaries accurately is shown in Figure 8.

Differences in drainage patterns were easily noted, a differentiation especially important in the region of alfisol soils, which exist in a temperate, moist climate, where profile differences among soil types are strongly influenced by profile drainage characteristics.

Because soil classification systems worldwide are based on characteristics of the total soil profile and not just on the A horizon (generally the surface soil), there has been much skepticism about the possibility of mapping soils from surface spectral properties. The Clinton County studies provided such good correlations between certain spectral boundaries and soil type boundaries that confidence began to grow regarding the use of surface soil characteristics as an indicator of the underlying horizons.

A concurrent study by the SCS and LARS of Chariton County, Missouri, helped greatly to resolve this problem.[18] In this study it was noted that when ancillary topographic data were used to stratify the scene into bottomlands, gently sloping uplands, and moderately sloping uplands, spectral data could be used to delineate soil types with a minimum of confusion. This is quite a logical approach because the surface soil characteristics of mature soils are strongly influenced by the macro climate of a region, whereas the characteristics of the subsoils are strongly influenced by the composition and history of the geologic parent material. In Chariton County the A horizons for each topographic subdivision were spectrally similar, having developed in the same major soil zone, but the total profiles were different because of the added influences of geology and terrain. We have since found similar stratification of land areas through ancillary data based on geology, etc., a great aid in bringing soil spectral maps into closer concurrence with soil series maps. The Chariton County study led us to concentrate on perfecting registration and overlay techniques. It also gave us additional experience in team work between soil scientists and data analysts.

Recently the SCS in Illinois has invited LARS to aid in investigating ways of using LANDSAT data to expedite the soil survey of Ford County. Also, we have begun a cooperative program to research the efficacy of using satellite data in making surveys of sparsely populated arid lands in the west.

To date the leaders in the studies in Indiana have been extremely pleased. A federal administrative officer said to me, "Your technique has made a science from what was formerly an art." It is too early to pass definitive judgment on how universally helpful to soil surveys spectral data will be, but in Indiana there is marked optimism among soil scientists with Purdue and those with the SCS.

## *HINDSIGHT GAINED*

This progressive series of studies, which led step by step to the current very promising experimentation to improve the use of remotely sensed data as an aid in making soil surveys, demonstrates the value of melding people from complementary disciplines when developing applications of remote sensing. In this case, the close association of data analysts with soil scientists ensured successful progress in discovering ways to use satellite data to expedite soil surveying and mapping.

Experience has shown that more precision is possible using LANDSAT data than is necessary for most soil surveys being made at this time. Judgment based on experience is needed to fit the spectral detail to the needs of the kind of survey being undertaken.

Also, more research is needed to aid in the interpretation of data. Much work is needed to determine the influence of specific physical and chemical properties of soils, taken singly or in combination, on their spectral properties.

## *POTENTIAL FOR INFORMATION TRANSFER*

Because the SCS is a well organized, well supervised organization, the transfer of new technology helpful in the national soil survey takes place rapidly and efficiently. The same is true of the Statistical Reporting Service, USDA, and of the Bureau of Land Management (BLM), USDA.

Other means of extending knowledge of the expanding technology are the usual symposia, seminars, training sessions, scientific meetings and publications, all of which are utilized by technical people in the fields of remote sensing, agriculture, and land use. Technology transfer is a major part of the program at LARS, as well as at several other centers which concentrate on the applications of remote sensing.

## *REFERENCES*

[1]Ball, G. H, and D. J. Hall. 1965. *ISODATA, a Novel Method of Data Analysis and Pattern Classification.* Stanford Research Institute, Menlo Park, CA. 1-16.

[2]Baumgardner, M. F.; S. J. Kristof; C. J. Johannsen; and A. L. Zachary. 1970. "Effects of Organic Matter on the Multispectral Properties of Soils." Proc. Ind. Acad. Sci. 79:413-422.

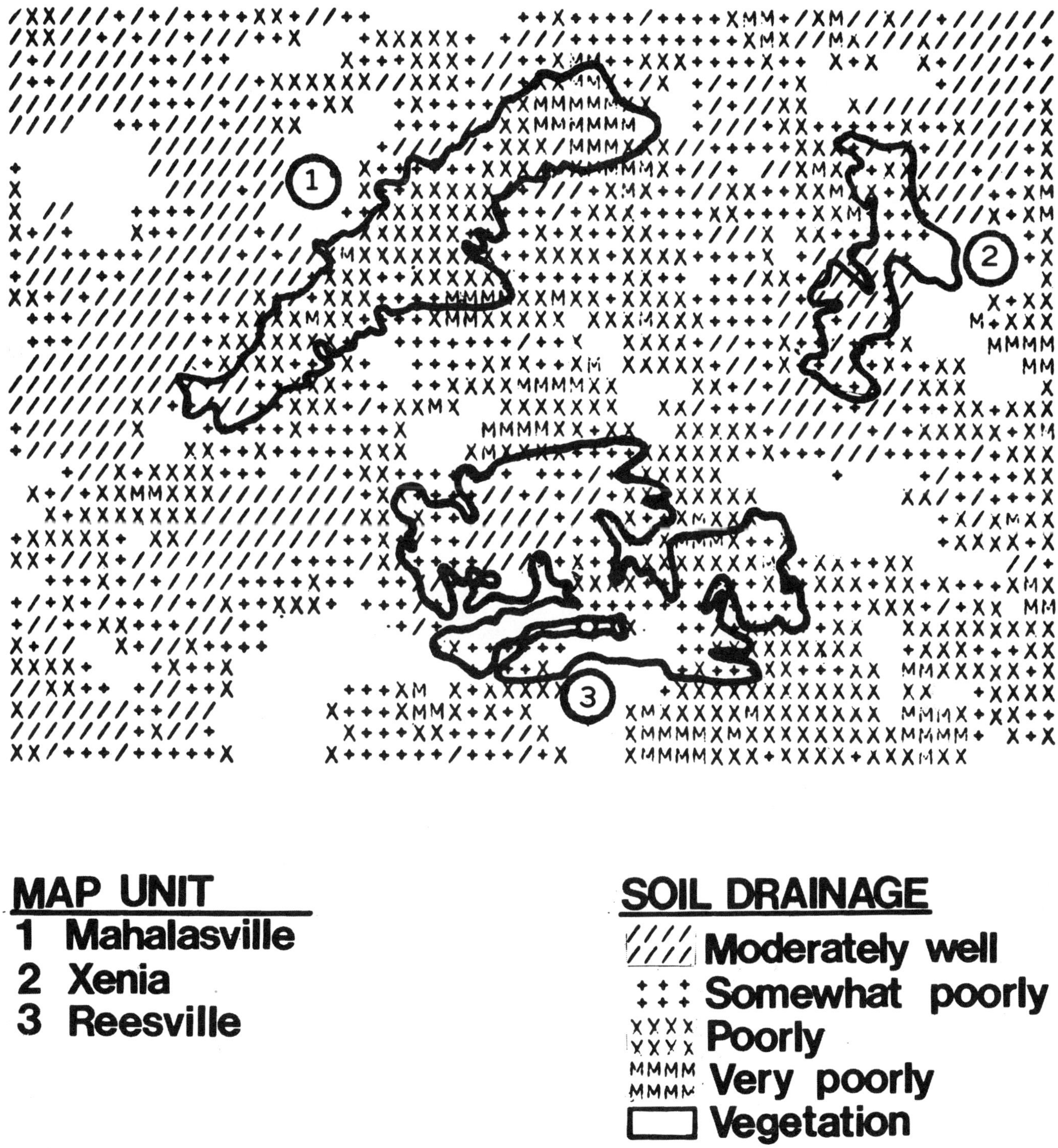


*Figure 8. Overlay of Conventionally Mapped Soil Boundaries over Spectral Map, Clinton County, Indiana.*

[3]Bushnell, T. M. 1929. "Aerial Photographs of Jennings County." Proc. Ind. Acad. Sci. 39:229-230.

[4]Cobb, W. B. 1922. Possibilities of the airplane in soil survey work. Amer. Soil Survey Assoc. Bull. 4:77-79.

[5]Condit, H. R. 1970. The spectral reflectance of American soils. Photogrammetric Eng. 36:955-966.

[6]Joerg, W. L. G. 1923. The use of airplane photography in city geography. Annals of Assoc. Amer. Geographers. 13:211.

[7]Johnson, G. R. and R. R. Platt. 1930. Peru from the Air. Special Publications 7, American Geographic Society, New York.

[8]Kellogg, C. E. 1938. Recent trends in soil classification. Soil Sci. Soc. Amer. Proc. 3:253-259.

[9]Kirschner, F. R., S. A. Kaminsky, R. A. Weismiller, H. R. Sinclair and E. J. Hinzel. 1977. Map unit composition assessment using drainage classes defined by Landsat data. Laboratory for Applications of Remote Sensing Information Note #092377. Purdue University, West Lafayette, IN.

[10]Kristof, S. J. 1971. Preliminary multispectral studies of soils. J.

Soil and Water Cons. 26:15-18.

[11]Kristof, S. J., M. F. Baumgardner and C. J. Johannsen. 1974. Spectral mapping of soil organic matter. ITC Journal. 4:479-489. Enschede, The Netherlands.

[12]Lee, W. T. 1922. The Face of the Earth as Seen from the Air. Special Publication 4. American Geographic Society, New York.

[13]Montgomery, O. L. and M. F. Baumgardner. 1974. The effects of the physical and chemical properties of soil on the spectral reflectance of soils. LARS Information Note #112674, Purdue University, West Lafayette, IN.

[14]Morain, S. A. 1974. Interpretation and Mapping of Natural Vegetation. Estes, J. E. and L. W. Senger (Eds.). Remote Sensing. Hamilton Pub. Co., Santa Barbara, CA 127-165.

[15]Munsell Color Company, Inc. 1954. Munsell Soil Color Charts. Baltimore, MD.

[16]Peterson, J. B., F. E. Goodrick, W. N. Melhorn. 1975. Delineation of the boundaries of a buried pre-glacial valley with Landsat-I data. LARS Information Note #100175, Purdue University, West Lafayette, IN.

[17]Phillips, T. L. ed. 1973. LARSYS User's Manual, Vol. 1, 2. Laboratory for Applications of Remote Sensing, Purdue University, West Lafayette, IN.

[18]Weismiller, R. A., I. D. Persinger and O. L. Montgomery. 1977. Soil inventory from digital analysis of satellite scanner and topographic data. Soil Sci. Soc. Amer. J. 41(6):1166-1170.

# *Habitat Characterization: A Production Mode Usage of a Geographic Information System*

*by Carl Reed*


## *ABSTRACT*

For the past three years the US Fish and Wildlife Service has been developing computer-based technologies to aid fish and wildlife management and protection. An integral part of this effort has been the development of a large-scale geographic information system called EIDA (Ecological Information and Data Analysis). The system comprises three major subsystems: Data Entry (WAMS), Data Base Management (MANAGE), and Geographic Data Display and Analysis (MOSS). The system, still under development, is being used on a production basis for habitat characterization. This paper will discuss both the system and its use for habitat characterization.[1]


## *INTRODUCTION*

In the western areas of the United States there are unique resource development problems. Water is scarce, and competition for its use intense. The development of energy resources such as coal and oil shale often creates conflicts with traditional land use practices. Such development can have dramatic impacts upon fragile, previously undisturbed ecosystems. Information on these impacts is needed for development decisions in the West, especially those that will affect fish and wildlife and their habitat. Consequently, for several years, the Western Energy and Land Use Team (WELUT) has been developing methodologies and tools that will allow for more rapid and accurate input into the development decision process. These have been termed Rapid Assessment Methodologies. This paper discusses one of the major WELUT efforts: the application of a large-scale geographic information system to habitat characterization. Habitat characterization is an analysis/modeling sequence that uses EIDA to produce measures of wildlife habitat quality for units of land 400 acres and up.

## *EIDA — BACKGROUND, ORGANIZATION, AND CURRENT STATUS*

EIDA is composed of three major components: data entry, data-base management, and display and analysis. Each of these major pieces is connected, or planned to be connected, with interface software (Figure 1). The characterization process requires using two of the major components. These are now discussed in greater detail.

### Data Entry

The US Fish and Wildlife Service has developed two programs for digitizing map-based information — MAPDRAW and WAMS. MAPDRAW is currently running on a mainframe that supports a time-share environment, for users who must access off-site mainframes to do their digitizing. WAMS is WELUT's in-house digitizing system that is currently running on a large mini-computer (Data General C330/350).[2] WAMS is being used in the habitat characterization process.

### WAMS

The WAMS (Wetlands Analytical Mapping System) was designed and produced by Autometric, Inc., under contract to the US Fish and Wildlife Service. The original impetus for the development was the National Wetlands Inventory, whose mandate was to delineate, classify, and map all wetlands in the U.S. To meet the objectives of this mandate, a sophisticated digitizing capability had to be developed.

A typical WAMS station consists of a digitizing table (ALTEK or Tektronix) or an analytical stereo-plotter (APPS), an alphanumeric terminal, and a Tektronix 4010 CRT. The host computer is a Data General Eclipse C330 with 512 bytes of memory. There are currently four stations interfaced to the Eclipse.

WAMS allows the user to digitize from maps or from photography. Digitizing is done in arc mode with the digitizer specifying starting and ending nodes and left, center, and right attributes. The actual segment digitizing may be done in point, curve, or stream mode. During digitizing, the operator may delete a segment or "backtrack" and delete a portion of a segment. As the operator is digitizing, the digitized data are displayed on the Tektronix 4010.

Once a map sheet has been digitized, a machine-based arc to polygon procedure is run. Whenever an error is encountered, informative diagnostics are issued and the operator can

[1]The opinions expressed in this paper are those of the author and do not necessarily reflect US Fish and Wildlife policy.

[2]Mention of brand names does not indicate endorsement.

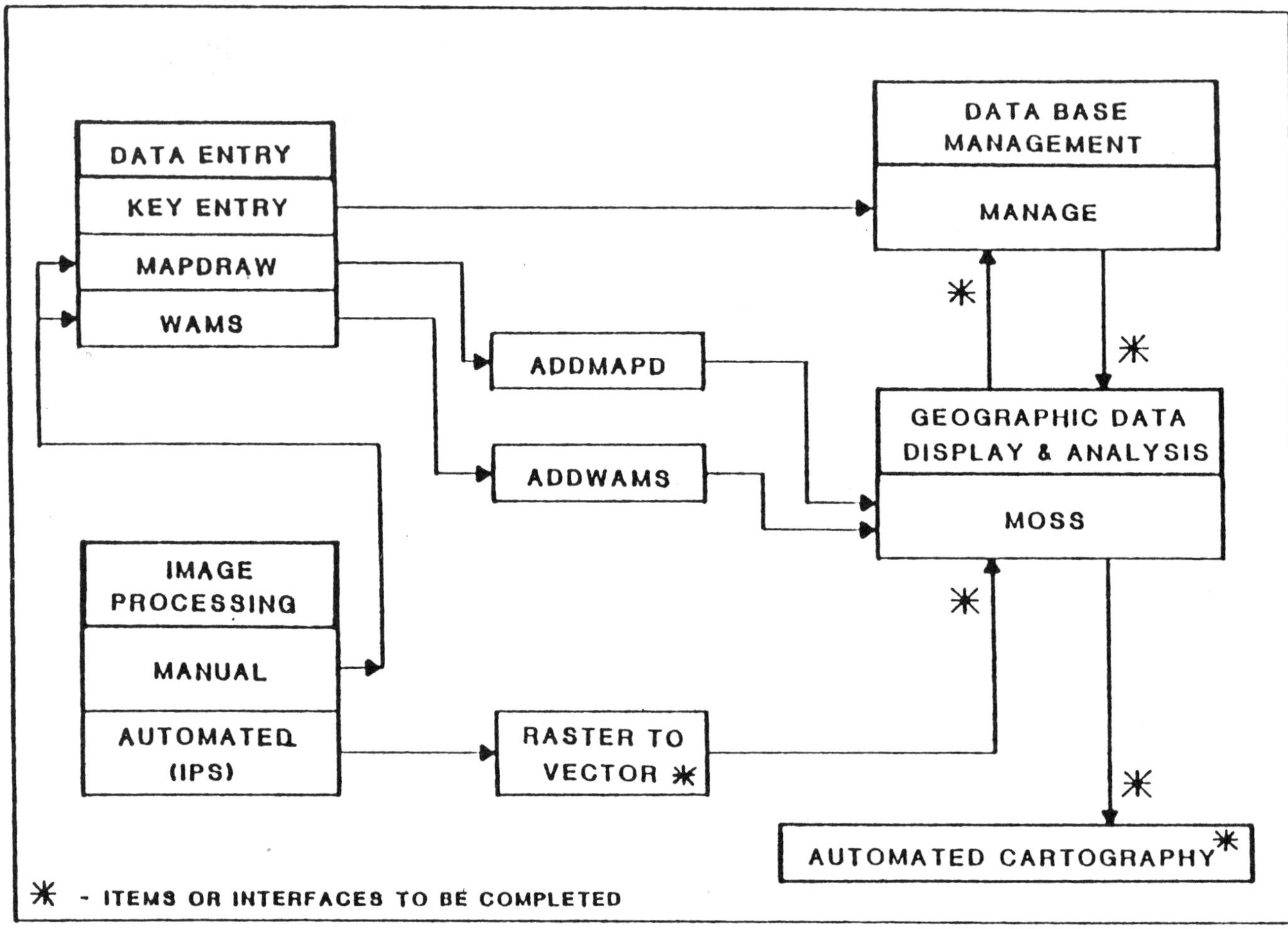


*Figure 1. Organization of EIDA*

edit the offending node or arc. After the map is verified, it can be test plotted on a Calcomp plotter. If the map seems clean, it is data-based for later use in joining edges between adjacent map sheets.

WAMS is now operational. Using WAMS, a skilled digitizer can produce between 100 and 200 edited polygons per day.

### Data Display and Analysis — MOSS

The first version of MOSS (Map Overlay and Statistical System) was developed under contract by the Federation of Rocky Mountain States for the US Fish and Wildlife Service (FWS). The current version, which is a complete revision of the contractor-supplied software, is now being maintained and further developed by FWS.

MOSS is an interactive FORTRAN program designed for the analysis and display of geographic data. These geographic data may be one of six types: point, line, polygon, raster, elevation point, and elevation grid. Functionally, MOSS has five groupings of operations. These are: (1) general purpose, (2) data base management, (3) display, (4) analysis, and (5) spatial retrieval. Table I presents a list of all the commands in MOSS and the basic function of each command.

Currently, MOSS is running on CDC equipment at Colorado State University, Boeing Commercial Services, and the Bureau of Reclamation in Denver. It is being converted to WELUT's Data General Eclipse. The functions necessary for characterization are already operational on the Eclipse. The remainder of the conversion is to be completed by August 30, 1979.

## *TERRESTRIAL CHARACTERIZATION*

In order to rapidly assess the impact of human development on wildlife habitat, a methodology had to be developed that both required minimal field work and was amenable to computerization. A group at WELUT has developed a process for evaluating regional wildlife habitat quality based on the analysis of recent color infrared aerial imagery, digitization, and calculation of habitat values. This process, called terrestrial habitat characterization, is a large effort. The initial coverage is the 8500 square miles of the Montana-Wyoming Regional Ecological Test Area. This is to be completed by October, 1979. An additional 20,000 square miles is being considered for characterization during 1980. As can be seen, this will be a production operation. Characterization will require most of the capabilities of EIDA for its successful completion. The characterization process is now described.

### Data Collection and Data Entry

Color infrared (CIR) aerial imagery at 1:31,680 scale is col-

Table I
Moss Commands

| | |
|---|---|
| **1. GENERAL PURPOSE** | |
| COST | printout runtime cost |
| FINISH | terminate MOSS run |
| HELP | Command tutorial |
| NEWS | print out latest MOSS news |
| LOCATE | point to screen location with cursor print out map coordinates for that point |
| QUERV | point with the cursor to a displayed point, line, or polygon and find out what it is |
| AUDIT | print out detailed summary of all items in a map |
| **2. DATA BASE** | |
| LIST | database browsing |
| SELECT | select all or any portion of a map for further display and analysis |
| ACTIVE | teller user what maps or parts of maps they have selected |
| FREE | kill an active map |
| STATUS | print out current status of a map |
| MERGE | merge 2-20 maps together into one map |
| POLYCELL | vector to raster conversion |
| SAVE | save an active map in user data-base |
| STUDYAREA | define arbitrary studyarea with cursor and clip source maps to generate new set of maps with data only in new study area |
| DUMP | dump the contents of a map file to the line printer |
| ADD | add a new map to the data-base |
| **3. DISPLAY** | |
| ERASE | erase screen |
| RESET | set virtual display window back to original |
| BLOWUP | define rectangle with cursor and set virtual window to it |
| PLOT | plot a map or set of maps |
| SHADE | crosshatching of polygons or cells |
| WINDOW | set virtual window to a map |
| LINE | plot line data in one of 18 symbologies |
| CELLPOLY | raster to vector (for display only) |
| TESTGRID | draw a grid over a polygon map |
| LEGEND | label a map on the CRT |
| WRITE | "write" a map to the plotter (vector data), to the printer (celldata), or to disk (transport data) |
| THREED | 3-D block diagrams or histograms |
| **4. ANALYSIS** | |
| DISTANCE | find distance between two points or along a line |
| COMPOSIT | arithmetic and logical cell compositing |
| LENGTH | length table for line data |
| PERIMETER | perimeter table for polygon data |
| AREA | area table for line data |
| FREQUENCY | frequency table for any vector data |
| GRID | evevation point to grid |
| SLOPE | slope calculations |
| ASPECT | aspect calculations |
| BUFFER | calculate buffer zone about point, line, or polygon data |
| OVERLAY | polygon overlay |
| STATISTICS | histograms, crosstabs, and descriptive |
| MODEL | executive a text file of MOSS commands |
| **5. SPATIAL RETRIEVAL** | |
| SIZE | select line and polygon data on basis of size criteria |
| PROXIMITY | proximity analysis |
| CONTIGUITY | juxtaposition |
| EDGE | edge calculations |

NSUMBS = 14

AREA SUMMARY FOR MAP BUCRSCSCT ACTIVE MAP NO. 2

| SUBJECT | AREA | FREQUENCY | PERCENT |
|---|---|---|---|
| 7.2 OP UPLAND MIX SHRUB/HERBAC | 11079.70 | 59 | 32.74 |
| 7.1 OP SAGE SHRUB/HERBACEOUS | 2170.81 | 14 | 6.42 |
| 4.6 OP UP MIX SHRUB/SCAT PON-J | 2109.00 | 20 | 6.23 |
| 4.4 OPEN UP MIX SHRUB/SCAT PON | 3382.73 | 39 | 10.00 |
| NC NON-IRRIGATED CROPLANDS | 2729.37 | 17 | 8.07 |
| 4.7 OP RIP DEC SHRUB/SCAT RIP | 576.97 | 7 | 1.71 |
| 3.1 OPEN PONDEROSA/HERBACEOUS | 1042.83 | 10 | 3.08 |
| 2.2 OPEN PON PINE/UP MIXED SHR | 8575.58 | 48 | 25.34 |
| 9.1 OP UP HERB/SCAT SAGEBRUSH | 588.51 | 6 | 1.74 |
| 1.1 CLOSED PONDEROSA PINE FOR | 186.59 | 6 | .55 |
| 7.3 OP RIP DEC SHRUB/HERBACEOU | 94.59 | 4 | .28 |
| 2.6 OPEN PON-JUN/UP MIXED SHRU | 1105.83 | 13 | 3.27 |
| 2.5 OPEN PON-JUNIPER/SAGEBRUSH | 175.32 | 4 | .52 |
| 4.1 OPEN SAGEBRUSH/SCAT PONDER | 21.73 | 1 | .06 |
| TOTAL (IN ACRES) | 33839.54 | 248 | 100.00 |

ENTER COMMAND

*Figure 2.*

lected. The CIR is used for the delineation of surface cover types, using a minimum mapping unit of 40 acres. The interpreted surface cover polygons are then transferred to 1:24,000 mylar overlays.[3] The interpretation/transfer process averages 1.5 man weeks per 7.5 minute quad. These mylar overlays are digitized and edited using WAMS. On the average, two days are required to digitize, edit, verify, and insert a quad into the MOSS data-base. The average 7.5 minute quad contains 200 polygons and about 30,000 coordinate pairs. Polygons range in complexity from 5 to 10,000 coordinate pairs. Some polygons have as many as 90 islands. The public land survey section boundaries for each quad are also being digitized. In total, 170 quads are to be digitized for the Montana-Wyoming test area. To date, about 130 quads have been digitized. We are running three shifts a day on three stations and expect to have the data digitized by the middle of July. The calculated data volume for the surface cover data set is about 45,000,000 characters (@6,000,000 coordinate pairs) representing 43,000 polygons.

Once a quad has been digitized, the data are transformed from the WAMS digitizing data-base structure to the MOSS map data-base structure, using a program called ADDWAMS. With MOSS, an area summary table and choropleth map are generated (Figures 2 and 3). This procedure serves two functions: provides a last visual verification of the data quality, and provides a quick indication of the complexity of the surface cover for that quad. MOSS then computes a polygon intersection between all the sections for a quad and all the surface cover for a quad.

The polygon intersection process generates a set of new maps in which the unique surface cover types are defined for each section (Figure 4). The average time for this procedure on the Eclipse is about one hour per quad, operating in a very busy environment. This information is then extracted from the MOSS data base and stored in what is called the Characterization Data Base. This step is necessary in order to make a single pass through all 8500 sections during the execution of the characterization models. Consequently, all the information for all the sections must be in one integrated structure.

After the characterization data-base has been built, any of the characterization models may be run against it. The output from these models is a number that indicates the quality of habitat for wildlife in general. The information is used by USFWS to evaluate each section in terms of suitability for coal mining (Figure 5). The final product of the characterization process will be a set of 1:100,000 maps scale in which each section is labeled and has the habitat value assigned to it.

## THE FUTURE AND CONCLUSION

This paper has presented a brief review of the US Fish and Wildlife Service's use of EIDA in a production mode for habitat characterization. The terrestrial habitat characterization effort

[3]There are plans to test the adequacy of using satellite data as the primary source for characterization in 1980. This process would use the automated image-processing capabilities being developed for EIDA.

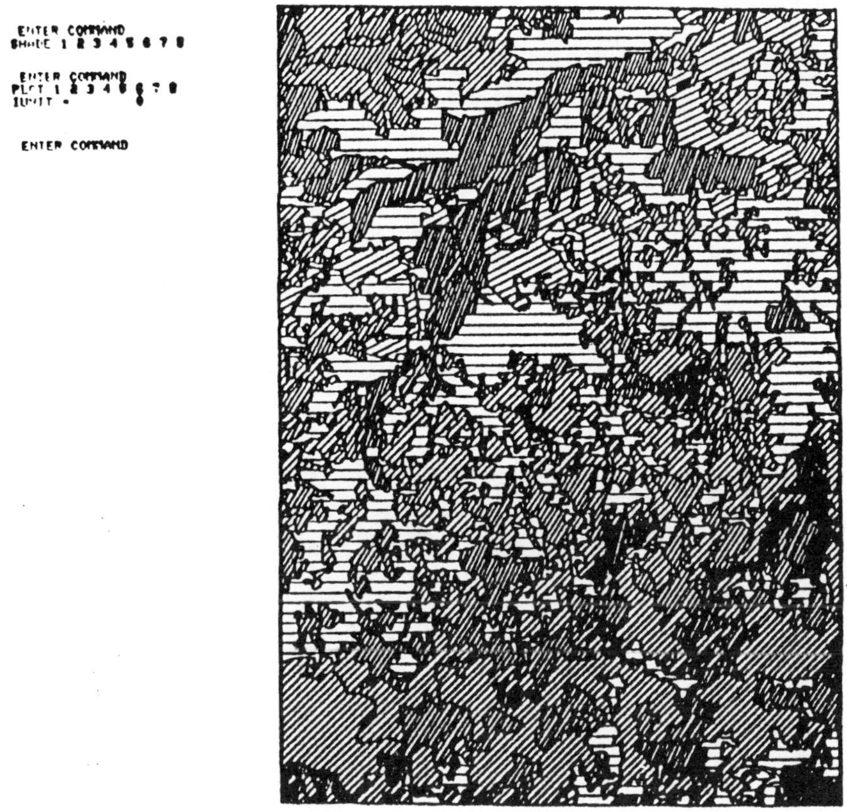


*Figure 3.*

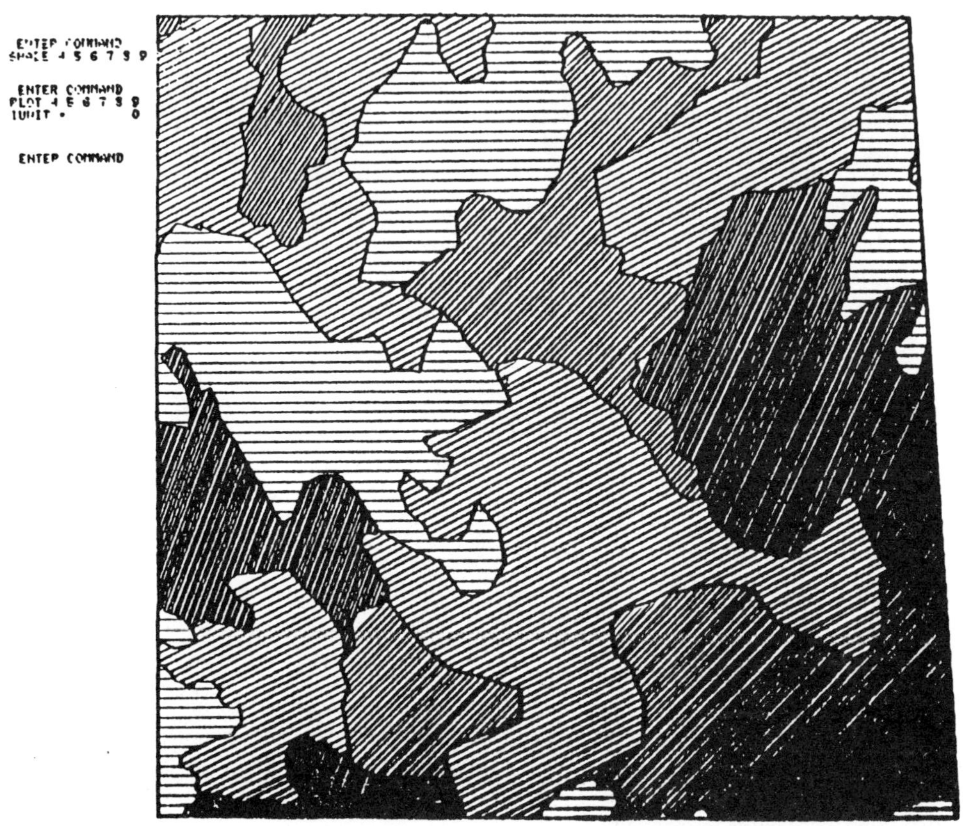


*Figure 4. One Section*

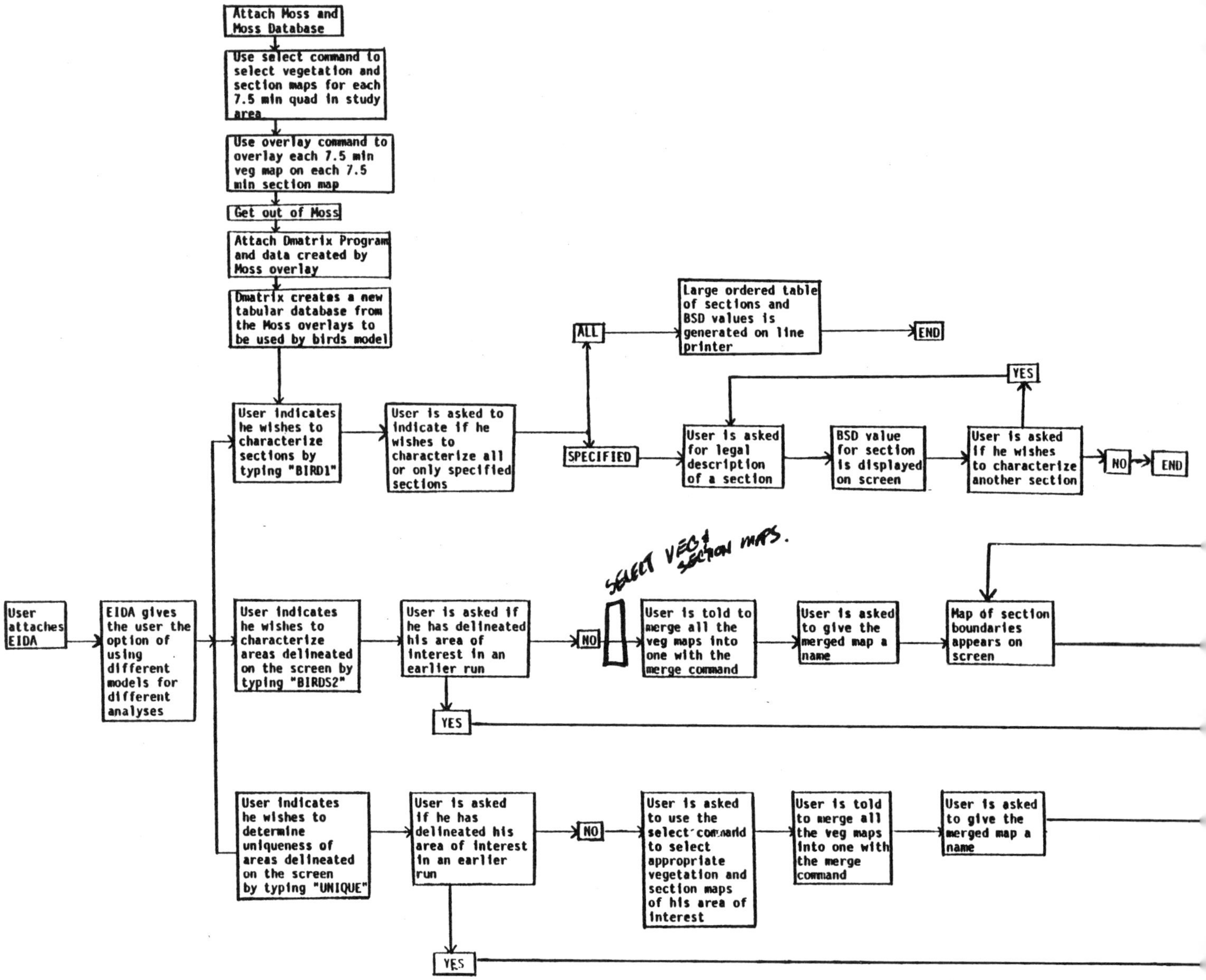


*Figure 5. General Schematic of the Terrestrial Wildlife Habitat Characterization Process with EIDA*

will be continuing through 1980. During this period an additional 20,000 square miles are to be mapped, digitized, processed, and modeled. to complete this operation, we are adding another CPU to our Eclipse, adding two more digitizing stations (to make a total of six) and adding several more characterization models. There will also be a full photo-interpretation staff and a map compilation staff.

Future analysis enhancements to EIDA include raster to polygon conversion, a cartographic subsystem of MOSS that will allow for the production of separates for color map production, and the addition of several commands to MOSS. The continuing development of the automated image processing is also planned. It is planned that these enhancements will be completed by 1981.

The FWS is a relative newcomer to the field of geographic information systems. However, manaagement support and a dedicated group of highly skilled technicians have been able to accomplish much in the past two years. In parallel to the technical developments, training has been ongoing. This training is viewed as a mechanism to spread the capabilities being developed by WELUT to the larger FWS community.

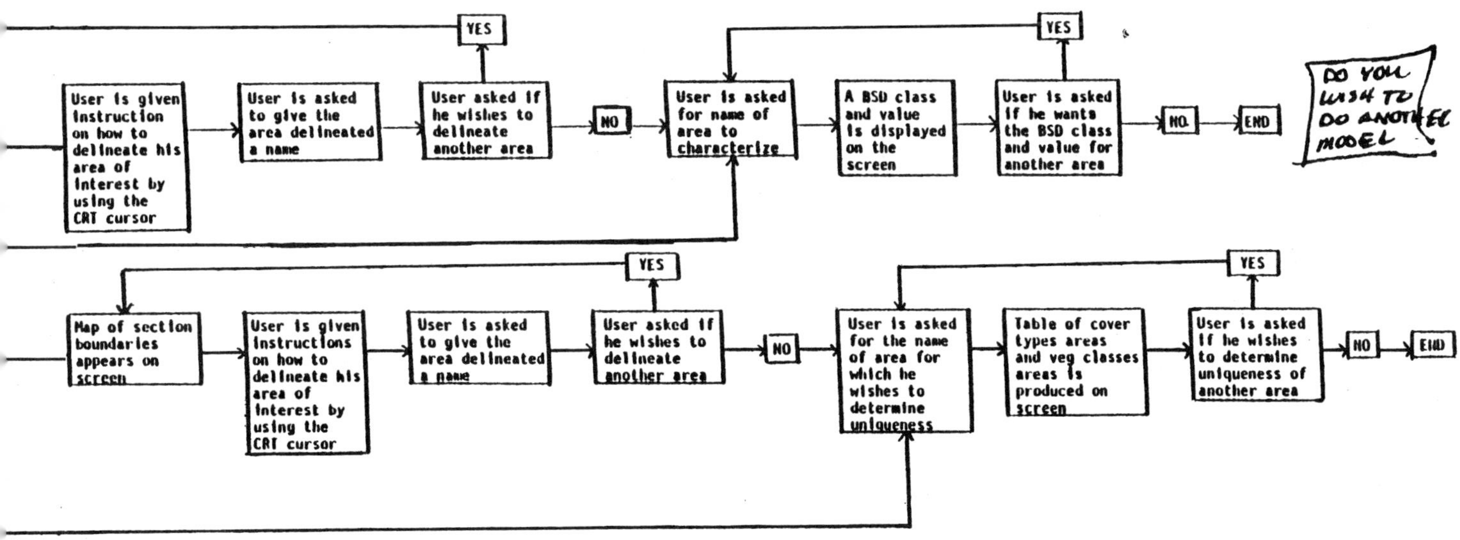


## *BIBIOGRAPHY*


Asherin, D.; H. Short; and J. Roelle. 1979. "Regional Evaluation of Wildlife Habitat Using Rapid Assessment Methodologies." presented at the Forty-fourth North American Wildlife and Natural Resources Conference, Toronto, Canada.

AUTOMETRIC, Inc., 1979. *The Wetlands Analytical Mapping System Users Manual.* Autometric, 5205 Leesburg Pike Suite 1308/Skyline 1, Falls Church, VA.

Gropper, J., and C. Reed, *MOSS Users Manual.* 1979, OBS, US Fish and Wildlife Service, Fort Collins, CO.

Reed C., and J. Gropper, *MOSS Systems Manual.* 1979. OBS, US Fish and Wildlife Service, Fort Collins, Co.

# *Studies of Offshore Petroleum Leases Using a Cartographic Data Base System*

*by John L. Sibert*

## *INTRODUCTION*

Since 1954 the US government has offered for lease over 3500 tracts in the outer continental shelf (OCS) region. Some 83% of these offerings are located in the Gulf of Mexico, making it by far the most significant offshore petroleum region. As part of an ongoing research and development program sponsored by the Conservation Division of the US Geological Survey (USGS), the Los Alamos Scientific Laboratory has responsibility for maintenance of a data base containing production, revenue, and bidding information for all offshore leases. The data base currently contains information for all tracts offered for sale from 1954 through 1977. (Offshore tracts are commonly referred to as offshore oil leases, although strictly speaking "tract" refers to the physical location while a "lease" is a legal agreement, in this case between the government and one or more oil companies, to allow use of that physical location.) Data for the 1978 sales are currently being added to the data base.

In addition to data items describing production, revenue, etc. the data base also contains the legal description of the location of each tract. The balance of this paper describes an effort to verify these descriptions using a cartographic data base system.

## *THE CARTOGRAPHIC DATA BASE SYSTEM*

The data base system used in this application is a prototype system developed at the Los Alamos Scientific Laboratory under contract with the USGS. The philosophy underlying the design of the system and the system itself have been described in some detail elsewhere.* However, a brief description of the system is necessary to understand how it is applied.

*R. L. Phillips, "A Query Language for a Network Data Base with Graphical Entities," *Computer Graphics,* Vol. 11, No. 2, (Summer 1977) 179-185.

R. L. Phillips and J. L. Sibert, "A Cartographic Query System for Management of Ofshore Leases," *Proceedings, Auto Carto III: International Symposium on Automation in Cartography,* San Francisco, (Jan. 1978).

The data base system has a network structure (Figure 1) making it possible to form queries based on combinations of items located throughout the data base. To make this powerful query capability accessible to the casual user, a simple query language has been developed allowing even the novice user to query the system without writing programs. In addition, and most significantly for this application, the system allows the production of color-coded thematic maps using a computer output microfilm device.

### Legal Description of Offshore Tracts

The precise location of an offshore tract is given by a legal description relating to the public land survey. The survey extends into the outer continental shelf region, which is divided into areas roughly analogous to counties on land. The areas are further subdivided into blocks, each of which has a unique (within an area) identifying number. The legal description specifies all or part of one or more blocks. By producing maps based on these descriptions, it is possible to verify the correctness of the legal description and of other data items as well.

## *DATA VERIFICATION PROCEDURE*

During data compilation a series of seven (four Louisiana, two Texas and one MAFLA) maps, with tracts color-coded by sale data, were produced by hand. In order to compare the contents of our data base with these originals, we produced a set of 143 color 35mm slides each portraying seven sales for a single area, with color-coding selected to approximate that of the originals. A color xerox copy was made of each slide to facilitate comparison with the original manually produced maps.

The xerox copies were then compared to the originals and every discrepancy between the two was recorded. When possible the OCS number (a code uniquely identifying each tract) was recorded. In some cases, the OCS number was not recorded on the original and it was necessary to use block and area designations. Along with the identification code, a brief description of the nature of the discrepancy was recorded. For example, if a tract appeared on the original but was missing from the slide, it would be recorded as "missing D.B." A summary of the results of this procedure is given in Table I.

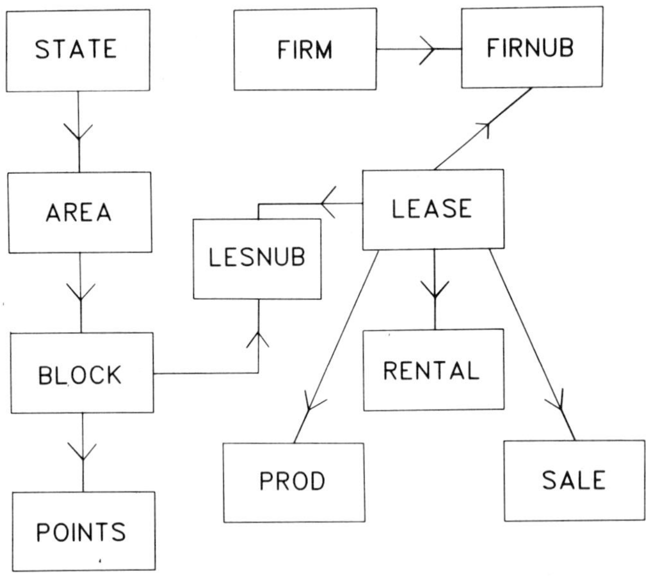


*Figure 1. Network Data Base Structure*

As can be seen from Table I, Louisiana has both the overwhelming majority of offshore tracts and, not surprisingly, of apparent errors as well. In addition, the proportion of tracts showing discrepancies is somewhat higher in Louisiana than in Texas. This may result from the higher complexity of the older land survey system in Louisiana, as well as the fact that Louisiana was involved in twice as many sales as Texas. MAFLA (an acronym for Mississippi, Alabama, Florida, Louisiana) would appear to have an even higher percentage of discrepancies, but ten of the thirteen turn out to have resulted from a single error in data loading.

The next step was to search the data base for each of the 290 tracts with discrepancies and try to determine the source of the difference. In some cases, supplementary maps were produced during this procedure. Figure 2 is an example of one of a series of such maps showing all tracts from a given sale. These maps were used to double-check some corrections which were sent to us during our procedure, as well as to see if comparison was truly facilitated by the use of color. Although no rigorous testing was done, it did appear much easier to detect discrepancies using the color-coded maps rather than the black and white ones.

During our search of the data base, a number of causes were found for the differences in the original and computer-generated maps. These are given in Table II along with their approximate percentage of occurrence. We found, as we expected, that most of the discrepancies could safely be attributed to common errors such as keypunch and recording errors. A few others were caused by programming bugs in the data base loading programs, while the remainder are to be referred back to the original data sources for reconciliation.

## *CONCLUSIONS*

The usefulness of any data base is directly related to the reliability of the data it contains. There are a number of methods both manual and automatic of verifying the accuracy of data. The more traditional methods, such as comparing lists of numbers or performing checks on the topological consistency of the data file, were inadequate in this instance. The ability to perform visual comparisons of data mapped from the data base had two primary advantages. First, the comparisons were done quickly. Less than one man-week of effort was necessary for the actual comparison. (Of course other parts of the procedure took considerably longer, but that would have been the case regardless of how the comparison was done.) Second, since it was relatively easy to see the discrepancies, the likelihood of some error being overlooked was greatly reduced. This is most important for data which are to be mapped, when a single error that might have been harmlessly submerged in the mass of data in a statistical analysis may show up clearly.

The combination of data base management and cartographic capabilities has much wider applicability than the data verification task described above. We have begun to use the system in statistical studies of the production and revenue data. Currently, we are also looking for further applications of the system in order to learn what future enhancements may be necessary. In the future, with constant advances in computer technology and the theory of how to use it, systems such as ours will become much more commonplace, providing an invaluable tool in management of our natural resources.

Table I
Occurrence of Discrepancies Between Computer-Generated and Hand-Drawn Maps

| | Texas | Louisiana | MAFLA | Total |
|---|---|---|---|---|
| Tracts | 728 | 2054 | 93 | 2875 |
| Discrepancies | 57 | 220 | 13 | 290 |
| Percentage | 7.83 | 10.71 | 13.98 | 10.09 |
| Areas | 27 | 43 | 13 | 83 |
| Discrepancies | 10 | 27 | 3 | 40 |
| Percentage | 37.04 | 62.79 | 23.08 | 48.19 |

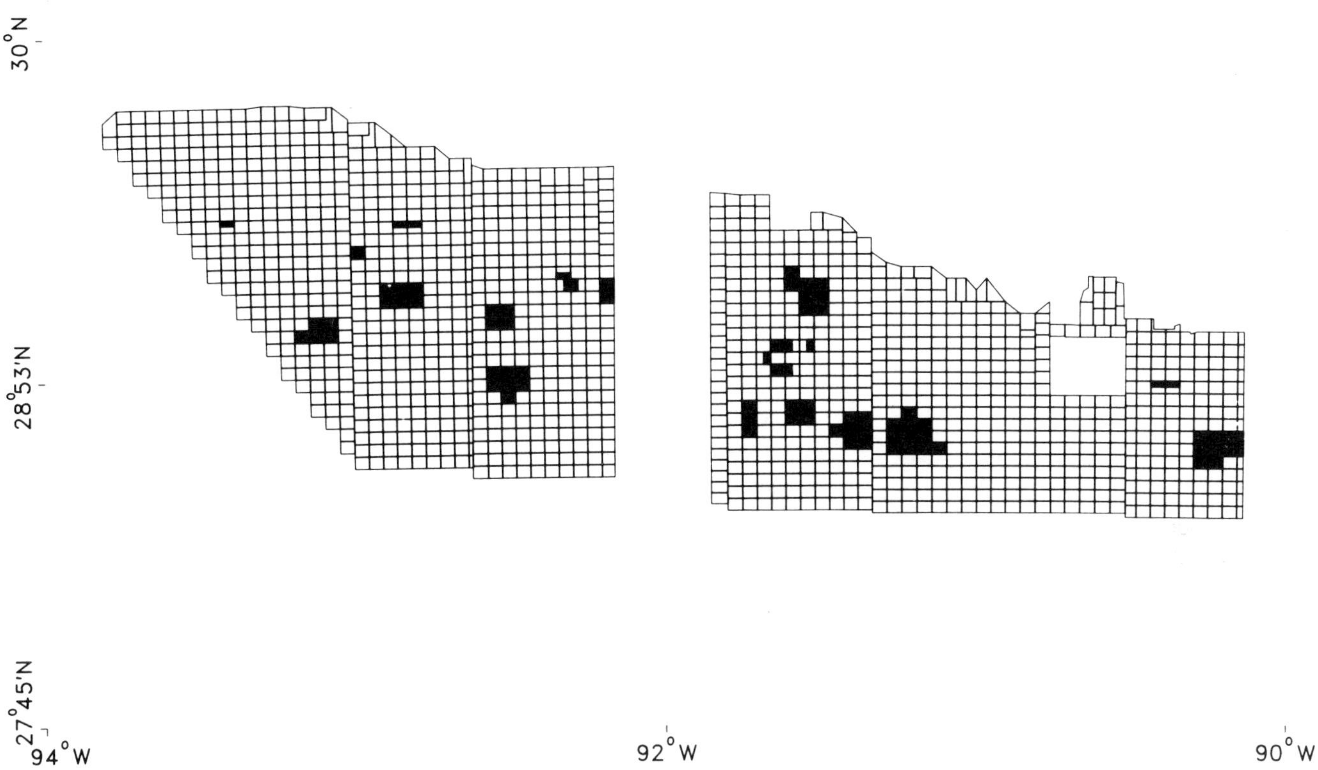


*Figure 2. Map Produced by Cartographic Data Base System — Lease Sale 10/13/1954*

Table II
Reasons for Discrepancies Between Computer Auto and Manual Maps

| Category | Percentage |
|---|---|
| 1. Bad Link Between Data Items* | 15 |
| 2. Date Difference Between Maps | 5 |
| 3. Erroneous Data Item* | 25 |
| 4. Discrepancies in Rejection and Reoffer Codes | 5 |
| 5. Missing Data Item* | 15 |
| 6. "Garbaged" Data Item** | 15 |
| 7. Bad AREA Record** | 15 |
| 8. Disagreement in Legal Description | 5 |

* Probable Data Entry Error
**Probable "Bug" Data-Base Build Program

## *BIBILIOGRAPHY*

CODASYL Data Base Task Group, April 1971 Report, ACM, New York.

Hershey, E. A. and P. W. Messink, "A Data Base Management System for PSA Based on DBTG 71." *ISDOS Working Paper No. 88,* University of Michigan (Jul. 1975).

Heindel, L. E. and J. T. Roberto, *LANG-PAK — An Interactive Language Design System.* American Elsevier, New York (1975).

Phillips, R. L., "A Query Language for a Network Data Base with Graphical Entities." *Computer Graphics,* Vol. 11, No. 2 (Summer 1977) 179-185.

Phillips, R. L. and J. L. Sibert, "A Cartographic Query System for Management of Offshore Leases." *Proceedings, Auto-Carto III: International Symposium on Automation in Cartography,* San Francisco, (Jan. 1978).

# The Use of Computer Graphics in Deer Habitat Evaluation

*by C. Dana Tomlin, Stephen H. Berwick, and Sandra M. Tomlin*

## PROBLEM

In 1977, the Yale School of Forestry and Environmental Studies initiated a study of the white-tailed deer *(Odocoileus virginianus)* population at the Yale Forest in northeastern Connecticut. The Yale Forest property comprises some 7,800 acres located in the towns of Union, Ashford, Eastford, and Woodstock, Connecticut, about 30 miles northeast of Hartford. The area is one of second growth forest, low hill topography, and distinctively rural New England character.

As one part of the deer population study, standard pellet group counts[1] were conducted during the spring of 1977 over an area of approximately one square mile. Within that area, five vegetative cover types were sampled to develop estimates of deer utilization. Among these cover types were red pine *(Pinus resinosa)*, white pine *(Pinus strobus)*, hemlock *(Tsuga canadensis)*, mixed hardwoods (predominantly *Acer saccharum* and *Betula allegheniensis*), and open field. Sampling sites were located along transects within each cover type. These were grouped at five sites per transect and five transects per cover type for a total of 125 sites. The first site of each transect was located randomly. Subsequent sites were then established at uniform distances along each transect line. At each sampling site, pellet groups were counted over a circular area of .01 acre or approximately 24 feet in diameter.

Analysis of the 1977 pellet group counts found statistically significant differences between cover types and indicated the following habitat preference ratings, expressed as estimates of the number of deer per square mile of each habitat type:

| | |
|---|---|
| hemlock | 124 |
| white pine | 71 |
| mixed hardwoods | 54 |
| red pine | 23 |
| open field | 1 |

Not yet considered at this point, however, (and often overlooked in studies of this sort) were effects due to the spatial nature of data being examined. Among the most important of these were, first, the inherent lack of statistical independence among observations made at neighboring locations, and secondly, the influence of specific spatial relationships between sampling sites and other locational features such as roads, water bodies, or areas of a particular vegetation type. It is this second group of spatial effects which has most recently been examined as part of the Yale Forest study and which is discussed below in terms of the computer graphic techniques employed.

The stated purpose of this phase of the deer habitat study was to examine, in a quick, flexible, and largely intuitive manner, the relationship between pellet group counts and certain factors associated wit the spatial context of each sampling site. Its more general purpose was, in fact, to demonstrate the speed, flexibility, and intuitive nature of one particular approach to that problem in the form of computer mapping. This effort was to be directed toward a local audience of student foresters generally unfamiliar with digital technology, uncomfortable with involved mathematics, and suspicious of anything not advertised in the L. L. Bean spring catalogue.

## METHODS

In keeping with these objectives, a decidedly small set of data was examined. Only two maps of the mile-square study area were encoded. The first was simply a collection of point values indicating pellet group counts for each of the 125 sampling sites. The second was a map of areas characterized by land cover types and site features. These included roads, structures, hemlock stands, red and white pine stands, hardwood areas, areas of mixed hardwood and softwood growth, open fields, and wetlands. This map is shown in Figure 1.

Both of these digital maps were derived from conventional maps compiled at a scale of 1:15,840 from field observations and aerial photographs. Pellet group counts were digitized as z coordinate values associated with x,y corrdinate locations. Land cover types and site features were defined by boundary lines digitized as sequences of x,y coordinate pairs.

These digitized coordinates were then used to create grid maps for subsequent analysis. A grid cell size of approximately 30 feet by 30 feet was chosen as one which would provide the resolution necessary to distinguish between sampling sites without requiring excessively large amounts of storage space. Each grid map contained 28,900 cells, dimensioned at 170 rows by 170 columns, and each cell was represented as a 16-bit integer. In the case of the pellet count map, each cell con-

[1] Pellet group counting is a method of estimating wildlife population density and forage use based on measurement of the periodic accumulation of fecal droppings at sampling sites within a given area. In this case, a defecation rate of 12.7 pellet groups per deer per day was assumed.

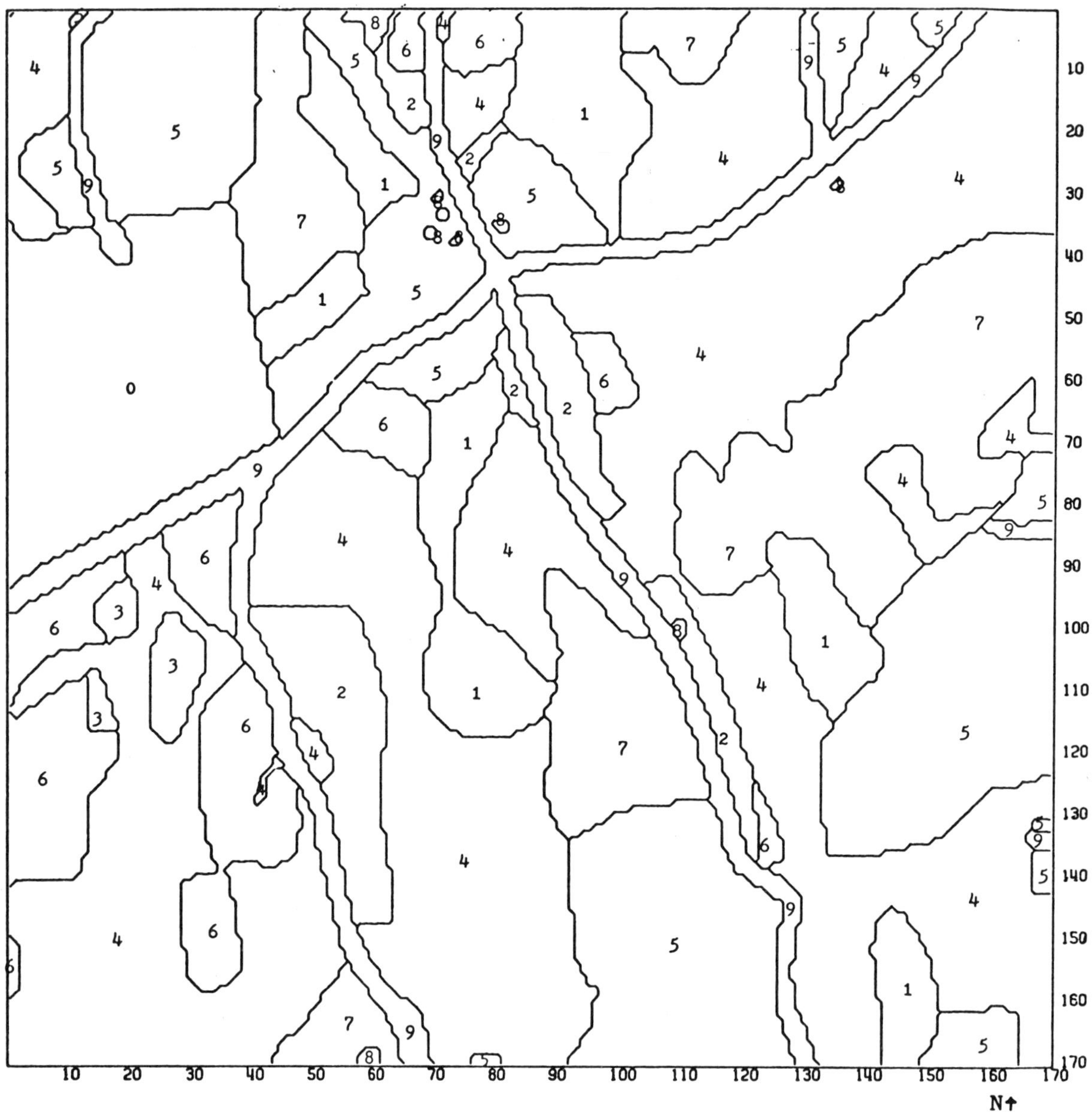


*Figure 1. Land Cover Types and Site Features: 0. Non-Yale land; 1. Wetland; 2. Open field; 3. Red pine; 4. Hardwoods; 5. Mixed hardwoods and softwoods; 6. White pine; 7. Hemlock; 8. Structure; 9. Road.*

taining the center of a sampling site (no cells contained more than one) was assigned the pellet count (x100) recorded at that site. All other cells were set to a value of zero. In the case of the land cover and site features map, values were assigned as follows:

0 land outside of the Yale Forest property
1 wet land
2 open field
3 red pine
4 hardwoods
5 mixed hardwoods and softwoods
6 white pine
7 hemlock
8 structure
9 road

Cells containing boundary lines, and therefore including more than one of the above categories, were assigned to whichever of those categories had the greatest value. The above values were initially assigned with this in mind such that each cell would be characterized according to its most "important" attribute.

Analyses of the spatial effect of land cover types and site features on deer pellet group counts generally followed a common procedure involving transformation of the cover type map to generate a new map of some particular spatial phenomenon; and, examination of pellet group counts, by cover type, within each region of this new map.

In one case, the cover type map was used to generate a map of "effective proximity" to roads and structures. This map is shown in Figure 2. Note here that "effective" proximity is clearly not measured in terms of Euclidean distance. Rather, it is defined in terms of units which may be consumed at rates which vary among cover types according to the degree to which those types either support or impede the movement of deer, i.e. units analogous to minutes of travel time or dollars of travel cost. In creating this map, it was assumed that areas of greater vegetation density (e.g. hemlock stands, hardwood stands, and areas of mixed hardwood and softwood growth), which generally offered better protection, were for that reason more conducive to deer movement. Open areas (e.g. wet lands and fields), on the other hand, were regarded as zones of higher "friction" relative to the movement of deer. It is for this reason that the proximity contours shown in Figure 2 are closer together in areas crossing wet lands and fields.

An alternative interpretation of "effective proximity" to roads and structures was also considered. Here, it was argued that it is not the movement of deer but the dissipation of noise, activity, and influence of man which should be measured and mapped. In this case, higher rather than lower "friction factors" were applied to areas of greater vegetation density.

Typical of the other factors examined in this manner were spatial effects such as the "local diversity" of cover types, the "narrowness" of forested areas, and the "shape complexity" of individual forest stands. A map of "local cover type diversity" is shown in Figure 3. Here, the value assigned to each grid cell indicates the number of different cover types occurring within a radius of approximately 25 meters around that cell. In creating the "forest narrowness" map, a value was assigned to each forest cell according to the shortest total distance from that cell to two diametrically opposing forest edges. "Shape complexity" values were computed by isolating individual forest stands (i.e. continguous areas of a single cover type) then characterizing each according to the ratio of its perimeter to the square root of its area.

Once mapped, these and other spatial factors such as stand size, proximity to hemlock stands, and visual exposure to roads and structures, were examined both individually and in weighted combination. In each case, categories of the spatial factor map, the land cover map, and the pellet count map were then compared on an overlay basis to generate cross-tabular statistics.

All of these analyses were performed in an informal and exploratory manner using packaged software in an interactive mode. The user-oriented nature and modular structure of that software proved to be particularly accommodating in this regard. Consider, for example, the following man-machine dialogue.

This is the command-response sequence generated in creating the effective proximity map shown in Figure 2. Here, "renumber," "spread," and "contour" are processing operations; "covermap," "development," "friction," and "proximitymap" are user-defined map file names; and "for," "assigning," "to," "thru," and "and" are parts of command-modifying phrases. "Relumber" is a mistake.

## RESOURCES

All software, hardware, manpower, and funding in support of this project were made available through the Yale School of Forestry and Environmental Studies. Data encoding, analysis, and display capabilities were provided through the use of software being developed as part of the Map Analysis Package (C.D. Tomlin, 1979). The Map Analysis Package is written in FORTRAN IV and may be used either interactively or in batch mode. At present, the package employs a grid-cell data structure for all analytic operations. Input data, however, may exist in the form of digitized points, lines, or polygons and output may be produced in the form of line-plotter graphics.

For this study, spatial coordinate data were encoded using a Talos RP648a digitizer and a GT44 graphics terminal linked to a PDP 11/40 processor. These coordinate files were then stored on disk packs accessible to Yale's IBM 370/158 mainframe computer. The 370, operating under IBM's Time-Sharing Option (TSO) and accessed through a Hewlett Packard 2648A graphics terminal, was used for all subsequent analyses. Preliminary graphics were generated by high-speed line printer while final maps were produced using a Calcomp 763 drum plotter.

Throughout this phase of the deer habitat study, slave labor was employed wherever possible. Digitizing was done by a first-year graduate student[2] of requisite steady hand and weak mind but without previous experience. The authors called upon similar traits in performing the analyses and preparing graphics. The analyses were performed at a total cost of about five CPU minutes.

[2]P.V. Brylski

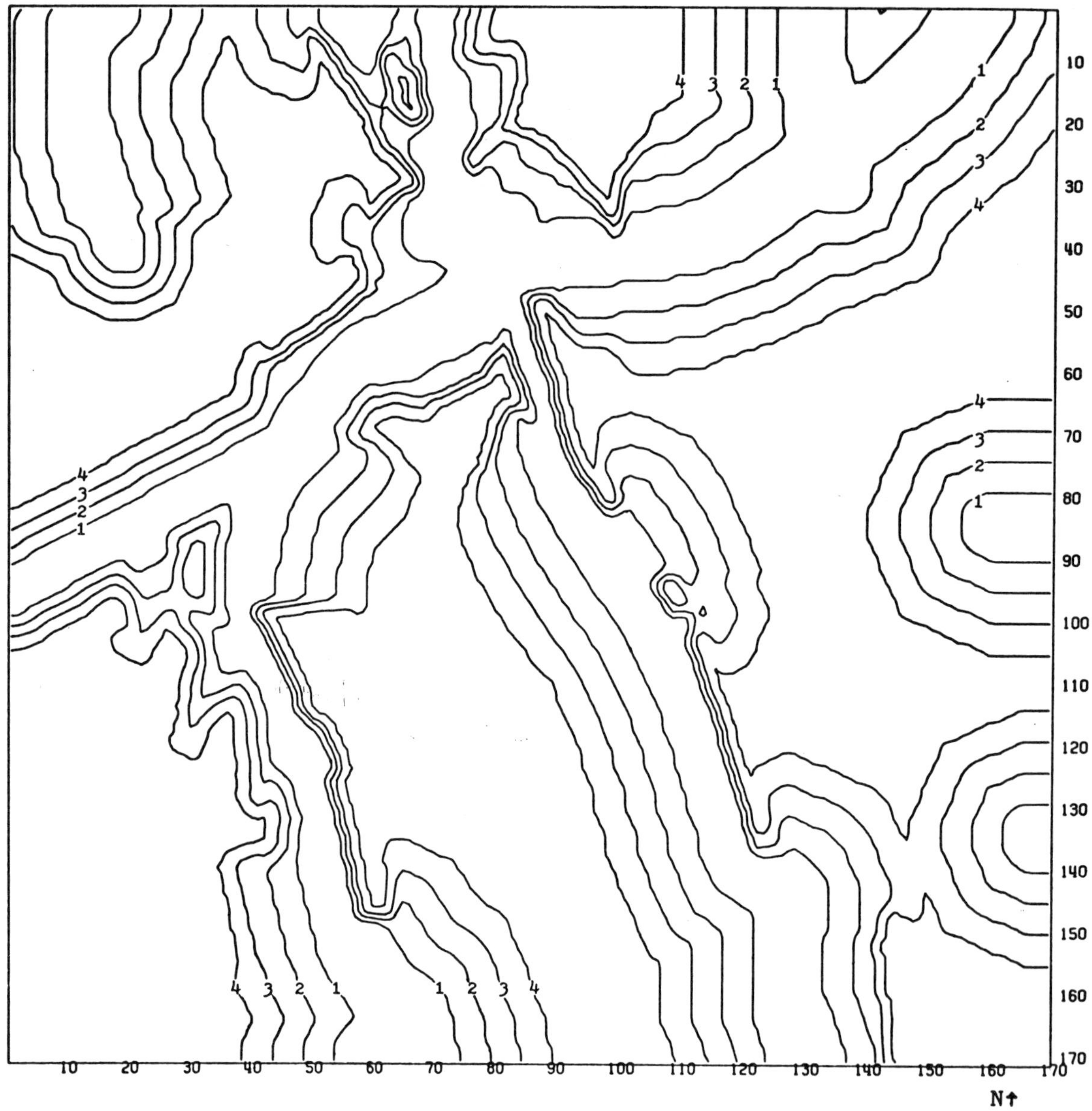


*Figure 2. Effective Proximity to Roads and Structures: 1. One unit away; 2. Two units away; 3. Three units away; 4. Four units away.*

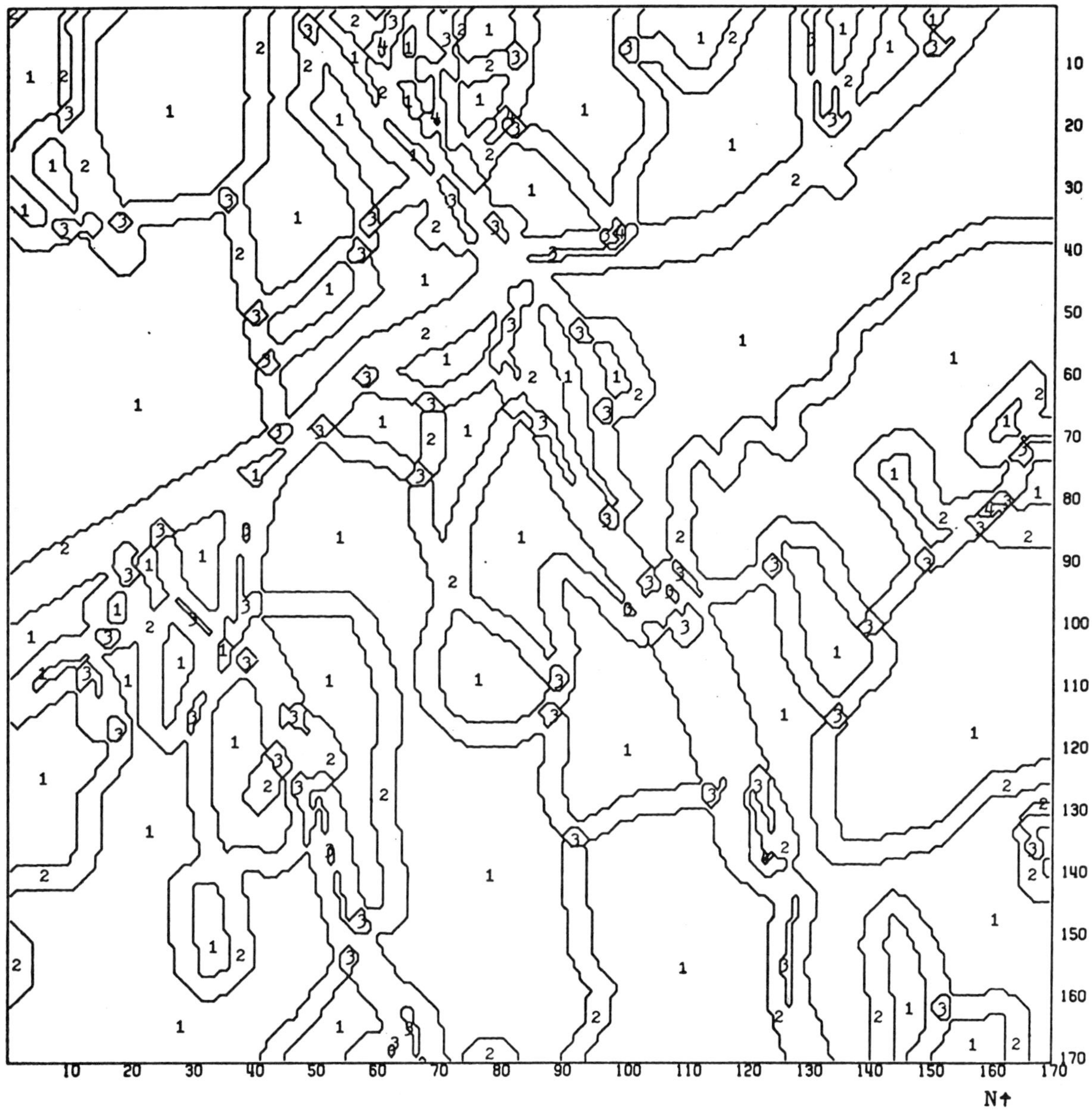

*Figure 3. Local Cover Type Diversity: 1. One cover type nearby; 2. Two cover types nearby; 3. Three cover types nearby; 4. Four cover types nearby.*

## PROBLEMS

Even in light of the limited scope and informal nature of this project, surprisingly few problems were encountered. By far the most costly and time-consuming aspects of the work were those associated with data encoding. This was largely due to specific problems in dealing with recently installed hardware.

Also somewhat frustrating was the fact that sampling sites were not uniformly spaced over the study area. As mentioned earlier, the pellet count data used in this study were not collected with spatial analysis in mind. While this presented no serious problems in terms of demonstrating the potential for applying computer mapping techniques, it did limit the ability to draw specific conclusions about the deer population in this particular area. The same was true in terms of the small number of independent variables considered. Most conspicuous by its absence was any consideration of topographic relief. Elevation data are now being encoded in hopes of incorporating factors such as slope, aspect, and cool air drainage.

## DOCUMENTATION

Two reports (Lyons, 1978 and 1979) are available describing the initial phases of the Yale Forest deer population study. Documentation on the more recent phases of that work involving computer graphics is forthcoming.

The Map Analysis Package is being developed as part of a doctoral dissertation on cartographic modeling techniques in environmental planning. A user's guide to the package is presently available in draft form. The Map Analysis Package approach to cartographic modeling is also supported by a graduate level course (Berry and Tomlin, 1979) and a series of user-oriented workshops sponsored by the Yale School of Forestry and Environmental Studies under the direction of Joseph K. Berry.

## GRAPHICS

The intuitive nature of graphics in general and the speed and flexibility of computer graphics in particular were found to be of critical importance to this project. This was true not only in the presentation of results but also in the formulation of the analyses themselves.

Examples of graphics produced are presented in Figures 1 through 3. All are grid maps reproduced in line-plotter form. Column/row reference numbers are indicated at the bottom and to the right of each map. Figure 2 was created by tracing contour lines over the three dimensional surface represented by effective proximity values. Figures 1 and 3 were created by tracing boundaries between cells of different values then rounding off right angles with diagonal lines. Labeling in each case was done by hand.

## RESULTS

Among the clearest findings of this project regarding the deer population in the Yale Forest study area were a mild direct relationship between pellet group count and local diversity of land cover types; and a mild inverse relationship between pellet group count and visual exposure to roads and structures.

In light of the lack of spatial distribution among sampling sites and the small number of factors considered, however, the significance of these results was felt to be questionable at best.

Despite this, the potential for applying computer graphic techniques in this area was clearly demonstrated. To the extent that this exercise might be characterized as a tool in search of an application, it did succeed in finding one.

## REACTION

Local reaction to the substantive content of this phase of the deer population study has focused with appropriate skepticism on the informality of the analyses performed. Reaction to the general approach demonstrated by this work, however, has been very enthusiastic. Particularly encouraging is the variety of projects making use of the Map Analysis Package at the University of Connecticut and at Yale. Among these are studies of bowhead whale populations in the north Pacific (Friedlander, 1979); airborne lead transport in Idaho (Von Lindern, 1979); spruce budworm infestation in Maine (Thorpe, 1979); the effect of microtopography on vegetation in New Hampshire (Gemborys, 1979); and watershed management (S.M. Tomlin, 1979), highway corridor siting (Kennard, Lefor, and Civco, 1979), and land use (Berry et al., 1979) in Connecticut. Developmental versions of this particular software are also now being used by several other universities and environmental consulting firms. Favorable reaction to the Map Analysis Package has generally focused on its intuitive command language, its range of modeling capabilities, and its organization as a series of independent "building block" operations.

## HINDSIGHT

In retrospect, several points have become increasingly clear. Had the initial phases of the deer habitat study anticipated the present concern for spatial effects, an alternative sampling strategy would almost certainly have been chosen. A larger set of data would also have been considered.

In terms of its ulterior motives in demonstrating the wonders of digital cartography, however, the simplicity of this study has proved to be one of its major assets. This has been the case not only in terms of software accessibility and analytic informality but also in terms of the conceptual simplicity of the grid-cell data structure.

## TRANSFERABILITY

Most of the data-processing capabilities used in this study can be found in many of the increasingly large number of packaged geographic information systems currently available. Experience at Yale and elsewhere suggests that such capabilities, if packaged in a sufficiently generalized manner, can accommodate a surprisingly wide range of applications.

Versions of the Map Analysis Package itself have so far been implemented on IBM, DEC, and HP computers. A CDC version may also be released. The package will soon be made generally available through Yale University.

### *BIBLIOGRAPHY*

Berry, J.K. and C.D. Tomlin, 1979. "An Academic Approach to Cartographic Modeling in Natural Resource Management." Second International User's Conference on Computer Mapping Hardware, Software, and Data-Bases.

Berry, J.K.; C.D. Tomlin; A. Friedlander; A. Howard; P. Korotky; P. Lewis; T. Niemann; J. Sailor; P. Scanlon; J. Tam; J. Thorpe; and I. Von Lindern, 1979. "Guilford Project Report: Cartographic Analysis of Residential Development Plans and Identification of Potential Landfill Sites." Yale School of Forestry and Environmental Studies.

Friedlander, A.I., 1979. "Analysis of Spatial Distribution of the Western Arctic Bowhead Whale (*Balaena mysticetus*) Population: Computer Mapping." Yale School of Forestry and Environmental Studies.

Gemborys, S.R. 1979. "Tree Distribution in Relation to Microtopography: The Bowl, Wonalancet, New Hampshire." Harvard Forest, Harvard University.

Kennard, W.C.; M.W. Lefor; and D.L. Civco, 1979. "Computer-Assisted Analysis of Earth Resources Information." University of Connecticut Department of Natural Resources Conservation.

Lyons, J.R., 1978. "The Status of Research on the Yale Forest, Tolland and Windham Counties, Connecticut." Yale School of Forestry and Environmental Studies.

Lyons, J.R., 1979. "Union Forest Deer Study: Report 2." Yale School of Forestry and Environmental Studies.

Thorpe, J.P., 1979. "Site Factors Affecting Spruce Budworm Infestation in Maine." Doctoral dissertation in preparation, Yale School of Forestry and Environmental Studies.

Tomlin, C.D. 1979. "Cartographic Modeling Techniques in Environmental Planning." Doctoral dissertation in preparation, Yale School of Forestry and Environmental Studies.

Tomlin, S.M. 1979. "A Digital Cartographic Data-Base for the Holcomb Farm Watershed." University of Connecticut Department of Natural Resources Conservation.

Von Lindern, I.H. 1979. "An Empirical Air Quality Model for Heavy Metal Contamination Surrounding a Primary Lead Smelter." Doctoral dissertation in preparation, Yale School of Forestry and Environmental Studies.